Cities of the World

A history in maps

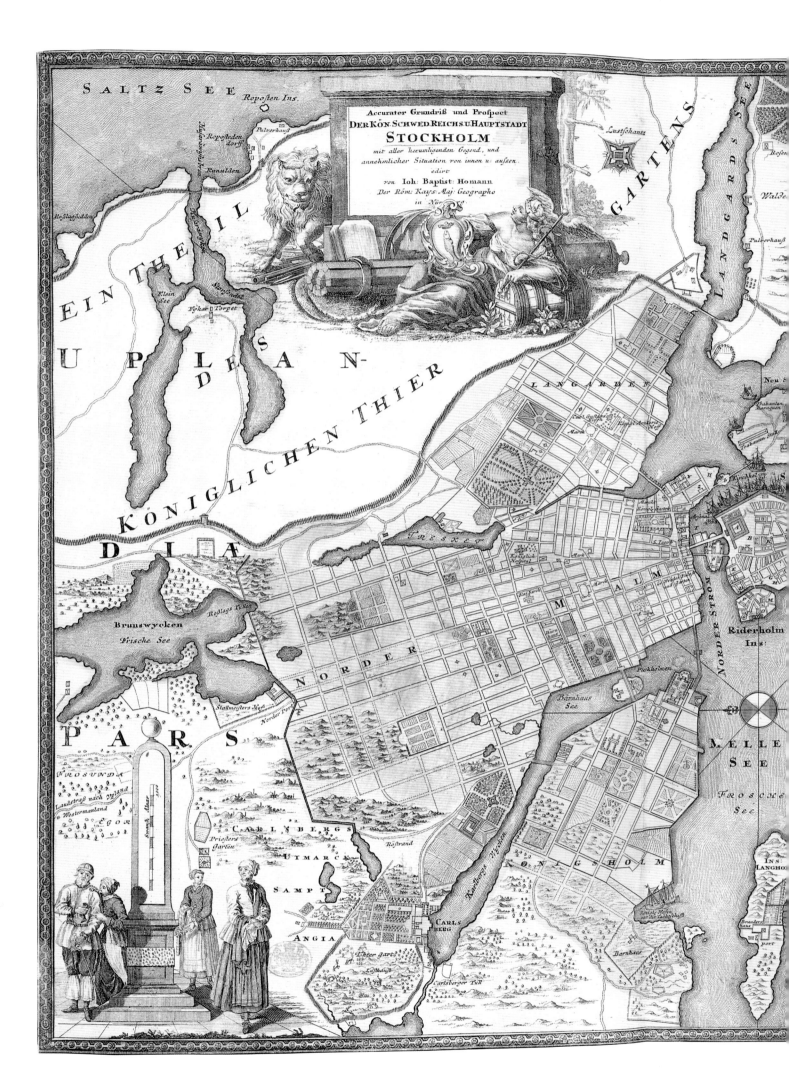

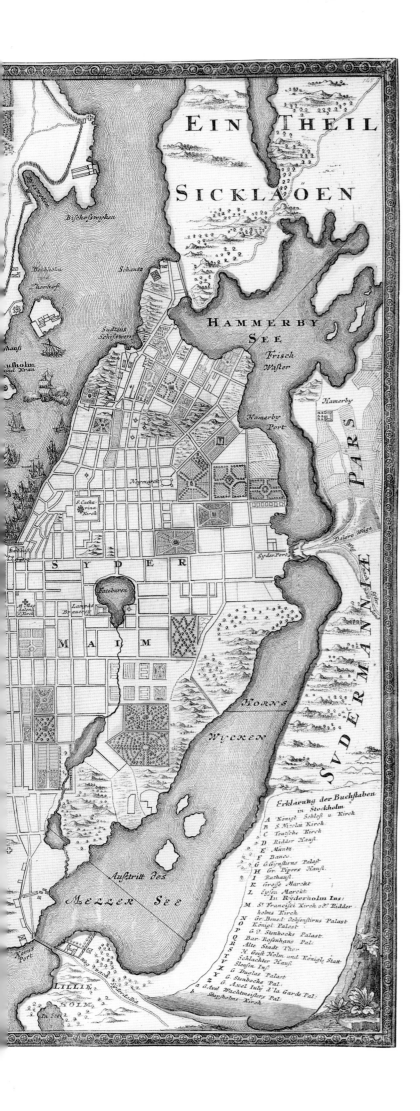

Cities of the World
A history in maps

Peter Whitfield

University of California Press
Berkeley   Los Angeles

University of California Press
Berkeley and Los Angeles, California

First published in 2005 by
The British Library
96 Euston Road
London NW1 2DB

Cataloguing-in-Publication data is on file with the Library of Congress
ISBN 0-520-24725-6

13 12 11 10 09 08 07 06 05
10 9 8 7 6 5 4 3 2 1

Designed and typeset in Mentor by Alison and Peter Guy
Map pages 6–7 by Cedric Knight
Printed in Hong Kong by South Sea International Press

Contents

8 Preface
9 Introduction: The City in History
27 Teotihuacan, a vanished city

CANADA

Quebec •

Chicago •   •Boston
• Salt Lake City   •New York
          Philadelphia•
          Washington •
• San Francisco   USA

                    AZILIA ▲ • Savannah
New Orleans •

          MEXICO

                    DOMINICAN
                    REPUBLIC
Mexico City • ▲TEOTIHUACAN   • Santo Domingo

          PERU   BRAZIL

          • Cuzco
                              • Brasília

                              • Rio de Janeiro

Tangier •

MOROCCO

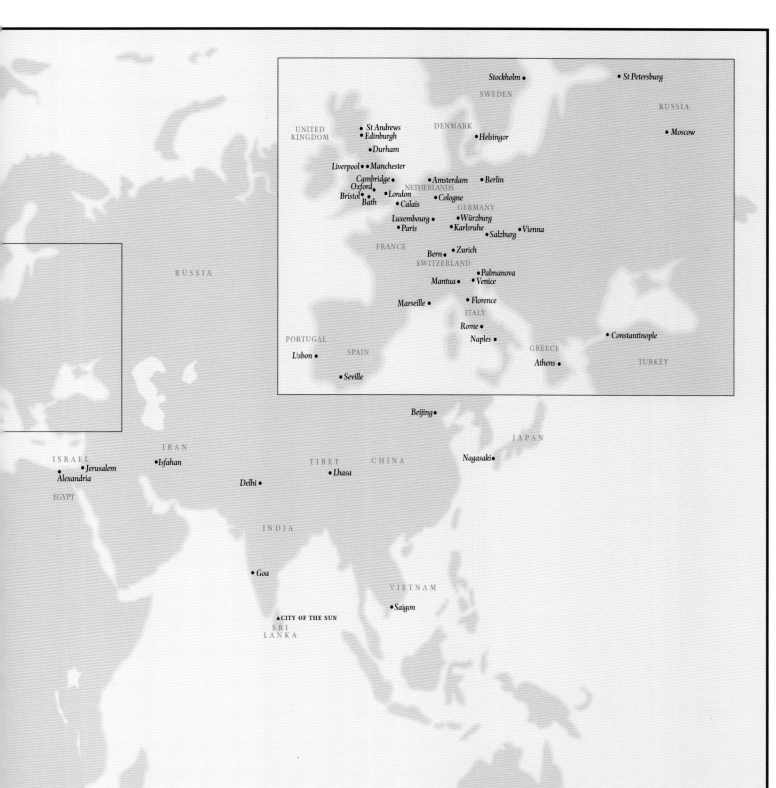

Stockholm •   • St Petersburg

SWEDEN

RUSSIA

UNITED
KINGDOM   • St Andrews   DENMARK   • Moscow
• Edinburgh   • Helsingor
• Durham

Liverpool • • Manchester

Cambridge •   • Amsterdam   • Berlin
Oxford •   NETHERLANDS
Bristol •   • London   • Cologne
Bath   • Calais   GERMANY
Luxembourg •   • Würzburg
• Paris   • Karlsruhe   • Vienna
FRANCE   Salzburg
Bern •   • Zurich
SWITZERLAND
• Palmanova
Mantua •   • Venice

Marseille •   • Florence
ITALY
Rome •
PORTUGAL   Naples •   • Constantinople
GREECE
Lisbon •   SPAIN   Athens •   TURKEY

• Seville

Beijing •

IRAN   JAPAN

ISRAEL   • Isfahan   TIBET   CHINA   Nagasaki •
• Jerusalem
Alexandria •   • Lhasa
EGYPT   Delhi •

INDIA

• Goa
VIETNAM

• Saigon

▲ CITY OF THE SUN
SRI
LANKA

AUSTRALIA

SOUTH AFRICA
• Cape Town

• Sydney

Cities of the World

Preface

'I am almost ashamed to say how tame and prosaic my dreams are grown. They are never romantic, seldom even rural. They are of architecture and of buildings - cities abroad which I have never seen, and hardly have hope to see. I have traversed, for the seeming length of a natural day, Rome, Amsterdam, Paris, Lisbon - their churches, palaces, squares, market-places, shops, suburbs, ruins, with an inexpressible sense of delight - a map-like distinctness of trace - and a daylight vividness of vision, that was all but being awake.'

Charles Lamb
Witches and Other Night Fears, 1823

In writing this book, I have, like Lamb, visited many of the cities in imagination only. This is not a travel book, but a historical survey of the form and the spirit of a number of famous cities. Those forms are at once apparent in the superb maps illustrated here. They show the historic heart of the cities: the ancient harbour, the hilltop fortress, the loop in the river, and the encircling walls, and they show the houses, churches and palaces which have been added over the centuries. Many of these historic maps have a pictorial quality which vanished long ago from the functional town-plan and, by showing the great buildings in elevation, they become architectural panoramas, capturing the richness of the urban fabric. Falda's Rome, Merian's Paris, Visscher's London or Revere's Boston offer us a god-like perspective on their subjects which no modern street-plan can equal. It is easy to see why the perspective view flourished for so long in the Renaissance period, before giving way to the scaled town-plan in the age of science.

The form of a city arises from its geographical setting and from its architecture. But where does its spirit arise from? Obviously from its citizens and from the historic events which have been acted out in its streets. In some subtle way, these personalities and events permeate the very air, and shape our perception of each individual city. As Henry James wrote of Rome, 'here was history, in the stones of the street and the atoms of the sunshine.' Cities are like great theatres, carrying the weight and the intensity of past triumphs. Oxford is so different from Rio, and Boston so different from Lhasa, that when we visit these cities *we* are different: we experience different emotions, we act different roles, and we become mirrors of the city. This explains our love affair with the world's great cities.

The sixty-four cities featured here were chosen not only because they are richly and beautifully illustrated, but also because they demonstrate this notion of spirit - an outward and inward uniqueness. This book attempts to record and explain this uniqueness. It shows how the city was linked to the birth and progress of civilisation itself, how it has acted as a focus for ideas and technologies, arts and sciences, and even for religious devotion. It shows the way that some cities have grown slowly into a haphazard, unplanned beauty, while others have been shaped or reshaped by the will of a masterful individual. Urban historians have long been searching in vain for the formula, the secret, of a great city, and this book too fails to find it. The only certainty is that the city's spiritual well-being is somehow linked to the physical form: beauty and order in the streets will inspire social harmony and individual peace - this is the ideal. That this ideal was achieved in the past but eludes us in the present is both a reproach and a challenge. The theme of this book, as of every other book on the history of the city, is to ask whether mankind can now control, civilise and enrich the urban environment which he has created for himself.

Introduction: The City in History

For most of us the city is the vessel that both contains and shapes our lives. Historically it has been the focus of all the ideas and energies which have guided civilisation. Nature was the environment into which mankind was born, but the city is the environment which he has created for himself, the arena in which he interacts with his fellow men. Man's troubled relationship with his cities, above all in the last two centuries, confronts him with this fundamental problem: can he control his own destiny, or must he surrender to the impersonal forces which he has set in motion? Those forces are social, economic and, increasingly, technological. The very word 'city' is related to the words 'civility' and 'civilisation', but that link has to be constantly questioned, restored and sustained. The city, majestic and squalid, intimate and impersonal, creative and cancerous, is the image of man himself, with all his vices and imperfections.

In the earliest phase of human culture, movement and settlement became twin poles about which tribal life revolved - the tension between security and mobility. The spur to movement was the perpetual quest for food, for the hunting and gathering of food could sustain perhaps a dozen people per square mile, and an ample, renewable food supply was therefore a precondition for settled existence and civilisation. In order to progress beyond biological survival, a surplus of food and ways to contain it were both required. This surplus could then support activities other than food-production: builders, craftsmen, priests, legislators, soldiers, and those with leisure enough to explore the intellectual realm. It is possible that the existence of a leisured class who can think, plan, invent and criticise is the hallmark of the urban culture which we call 'civilisation'. What is certain is that all these diverse activities combined to build up a conscious social life, richer and more complex than anything that had gone before: in some mysterious way the city acquired a soul.

This surplus of food came into being in the ancient Near East around 10,000 years ago, with the birth of agriculture - the cultivation of crops and the domestication of animals - and this swung the balance from a nomadic way of life towards settlement, to the camp, the shrine and the village. How and why these elements came together and evolved into the city in certain regions and not in others, no one knows. In Mesopotamia, Egypt, the Indus Valley, China and Meso-America, some cities were already old by 2000 BC, while in Europe, Africa, North America and most of Asia, the village endured. The cause of the transition was certainly political: the concentration of power in the hands of kings, who were able to centralise and direct the energies of the community. Kingship was vitally important in the early history of cities, and remained so for thousands of years. The hunter with his skills and his weapons turned his attention to the domination of other human beings, becoming their king, their warlord and their law lord. In his hands lay the disposition of the all-important food surpluses, and the power to initiate public building projects: city walls, palaces and temples. The building of the first cities in Mesopotamia and Egypt around 5,000 years ago utilised the first real machine, the human machine, composed of thousands of individuals working with a common purpose, like components in a vast mechanism.

The city was a part of the great outburst of creative activity which gave birth at this time to the wheel, the plough, the loom, the sail, mathematics, calendars,

metallurgy, coinage, writing, written literature and law codes. This concentration of political and technical forces led to an implosion, in which more and more people and energies were drawn into the city, as if to a magnet, making it the setting for still further innovation. People came to cities then, as they still do, in search of wealth, position, learning or power. To govern this complex and evolving structure, political and legal powers developed and became concentrated in the hands of rulers and legislators. But where the village was clannish and restrictive, the city was open and fluid. The origin of the city coincides exactly with the origin of writing: the need for records, laws, calendars and contracts created a collective memory, which became the starting point of recorded history. With its administrators, law courts, tax-gatherers and military quarters, the city became an embryo state. It was also the repository of ideas, supporting a professional class of intellectuals in temples, schools, academies, libraries and observatories.

Mesopotamia was the birthplace of the city and also the source of the earliest series of city images. Here a king sits enthroned within his walled city while his humble subjects ply their trades. From Layard's *Monuments of Nineveh*, 1849.
The British Library, Cup 648.c.1, pl.77

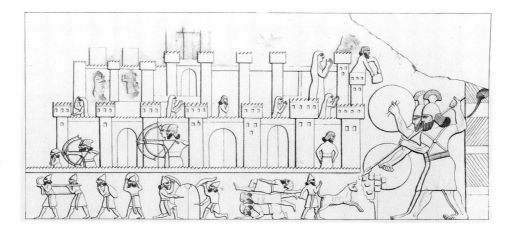

This intense specialisation of labour was characteristic of the city, making it the agent of man's self-transformation. Historically, the city and not the country has been the great source of ideas and of intellectual progress.

Where was the first city? Current evidence singles out Jericho and Catal Huyuk, as being settled by 8000 BC or even earlier, and many other similar sites may lie undiscovered beneath the sands. Catal Huyuk, in Anatolia, had an estimated population of 6,000, dwelling in mudbrick houses, while Jericho was even smaller, with little evidence of political or class structure to be found in either site. But cultic shrines were found in Catal Huyuk, and clear traces of metal-working and fabric-weaving, together with some remarkable wall-paintings including the first known depiction of a city. This is a fresco showing a panorama of terraced buildings, perhaps Catal Huyuk itself, with an active volcano in the background. There does not appear to have been any planning involved in Catal Huyuk; each building simply abuts the next in a maze of units. The crucial element in planning – public spaces and public architecture – is missing here, as it is in ancient Jericho.

On a very different scale were the dynastic cities of Mesopotamia, which emerged fully three or four thousand years later. The city of Ur may have had a population of 100,000, with temples and palaces, and there is clear evidence of a stratified society, codified laws, and a system of beliefs in which the king upheld order on earth as the gods did in the cosmos. The city had by then become a state, entire unto itself. The transition from village to city was not a mere shift in size, but in purpose and social organisation. Monumental architecture was set up deliberately to overawe both subjects and enemies, to proclaim the god-like magnificence of their rulers. These temples and ziggurats were an entirely new expression of human imagination and technical skill, something totally unknown in the village or in traditional arts.

But the city's wealth and power made it a natural target for aggressors from other city-states. 'In reality', Plato would later write, 'every city is in a natural state of war with every other city.' Hence the city, the container and engine of civilisation, became from the first dedicated to the destruction of other cities, and other civilisations. Hence the laments for ruined cities in ancient literature. In Mesopotamia, clusters of city-states evolved, which became merged by mutual warfare into regional empires - Sumerian, Akkadian, and then Babylonian and Assyrian. Within the city, order reigned in theory, but there was, and still is, struggle and competition built into city life: in courts, in markets, in business, in politics, in social life, and in the streets; and of course all these cities rested on the labour of large slave populations. Thus was born the sense of stress, conflict and fear that is endemic in urban life, and thus the consciousness of the rural–urban dichotomy, the idea of the era before cities as a time of innocence and a golden age. There was no single blueprint for these ancient cities, but one of the most intriguing of all symbols for a city was the Egyptian hieroglyph: a cross within a circle, symbolising the focal point where roads, men and merchandise met within the encircling walls. The market, the royal citadel and the religious shrine - these were the essential hallmarks of the city, in addition to the dwellings of the common people. In Egypt there was a unique variation, for each Pharaoh built his own capital city anew, which was also his tomb city, where thousands of workers laboured for years on his tomb or pyramid.

The emergence of cities in independent, widely separated cultures across the world, from Meso-potamia to Meso-America, suggests that the city is some-how a natural phase of human development, without being an inevitable one, since other cultures remained pre-urban for much longer. It seems hardly possible that the *idea* of the city could have spread and diffused, as the idea of the wheel or the sail did. Nevertheless, some historians have argued that certain features of Central American culture, including urbanisation, may indeed have arrived through the seaborne migration of peoples from the ancient Near East. For certain ancient cultures the city became a powerful symbol: a symbol of the highest achievement of human civilisation, an achievement which was possible only with the blessing of the gods, and one which was vulnerable to corruption and overthrow if divine favour was withdrawn. Marduk presided over Babylon, Athena over Athens, Jahweh over Jerusalem. Peace, law, good government and the common life were the ideals of city life, but the failure to realise these ideals led, in the case of Judaism, to the vision of a new, purified city, which would be restored in God's good time. In time, this dream of a new Jerusalem would become one of the most influential themes in Christian thought.

o o o

'Houses make a town, but it is the citizens who make a city'. In a few centuries the Greeks discovered more about man and nature than the Egyptians and Mesopotamians did in several millennia, and they made equally radical innovations in the forms of social life. The unique feature at the heart of the Greek *polis* was the citizen himself, with his democratic voice and his enjoyment of the public spaces of the city - the baths, the agora and the theatre, the latter becoming especially important as a public forum for ideas and entertainment, which had no parallel in the cities of Egypt or Mesopotamia. Aristotle defined the life of the Greek city as 'the common life for a noble end', and the *polis* itself became both an idea and an ideal, to be served and nurtured, while the quest for the ideal city now entered human thought and became eagerly debated - not the ideal in any physical or architectural sense, but in its laws and its social organisation. In fact, in contrast to our image of ancient Athens as an ideal, golden-age city, much of the place below the Acropolis was crowded, squalid and undistinguished. As one contemporary wrote around 300 BC: 'The road to Athens is a pleasant one, running between cultivated fields the whole way; but the city is dry and ill-supplied with water. The streets are nothing but miserable lanes, the houses mean, with a few better ones among them. On his first arrival, a stranger would hardly believe that this is the Athens of which he heard so much.' Athens was not synonymous with the Acropolis, which was a place apart, the shrine of the city's protective deity, on which were lavished the genius of the city's artists and the fervour of the people. The limitation of the *polis* ideal was its social exclusiveness: foreigners, women and slaves were all rigidly shut out from citizenship.

It was an essential part of the Greek concept of urban life that the city should remain of a manageable size, and this concept lay behind the Greeks' role as colonisers. As soon as a city's population grew beyond the optimum level, groups of citizens departed to found new cities, not only in Greece, but also throughout the Mediterranean region. This required a formal pattern of town planning that was readily replicated: a geometric structure, with streets intersecting at right-angles, was in many ways the opposite of natural community growth, yet it clearly suited colonisers planting new cities. Byzantium, Marseilles, Carthage, Syracuse, Trebizond and a hundred other cities were founded in this way as Greek settlements, and Alexander the Great was merely continuing this tradition in the cities which he planted throughout the Near East. These later cities of the Hellenistic age were in some ways very different from the classical *polis*, for this was a new political era, and the city was now shaped to be a showplace where the power of the ruling dynasty was on display, in the monumental architecture and wide boulevards - a theme that would appear again and again in city planning. Alexandria was consciously planned to display political and architectural magnificence in a way that Athens had never been.

The legacy of Greece as a city-builder passed to Rome, but in Rome's case there occurred a unique development in which a single city became a world empire, which extended its power by planting new cities, like seeds of its own culture, throughout Europe and the Mediterranean. It was the Romans who carried the city to the pastoral regions north of the Alps, where none had existed before. These Roman cities acted as centres of civilisation, and also

The Egyptian hieroglyph for 'city' symbolising a crossroads encircled by a wall. From L. Benevolo's *The History of the City*, 1980.
The British Library, L.42/777

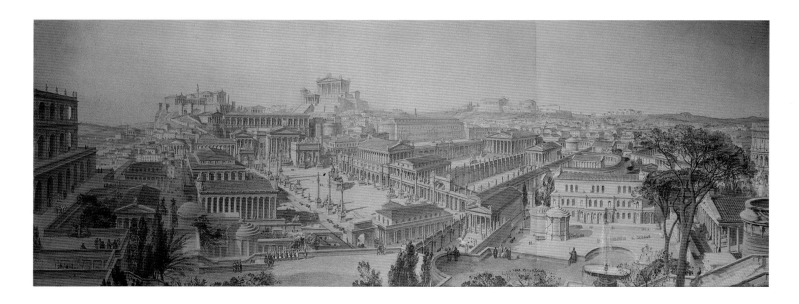

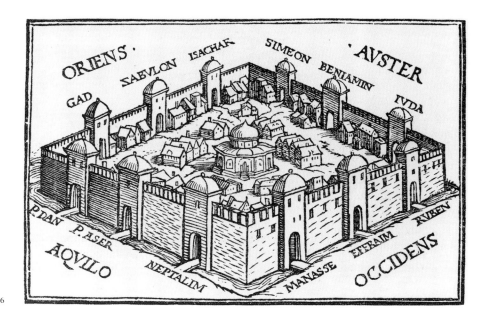

Above: 'Rome Restored' by Arthur Ashpitel, a nineteenth-century painting which brings to life the forum, the senate and the temples of ancient Rome.
Victoria and Albert Museum/ www.bridgeman.co.uk

Right: Jerusalem, the most frequently portrayed city in western art, from Holbein's Bible: it was symmetrical and orderly, walled against its enemies, with the temple at its heart, the image of the perfect, theocratic community.
The British Library, 012202.eee.16

The legacy of Rome in the Dark Ages: the arena at Arles was transformed into a small, fortified town.
The British Library, 10167.aa.37

as instruments of military and bureaucratic control. Their uniformity of purpose was reflected in the typical structure-plan of the Roman colonial city, built as a rectangle within protective walls, with two thoroughfares leading inwards from the gates. One street, called the *cardo*, ran north–south, and the other, the *decumannus*, ran east–west, meeting in the centre, where a forum, market, temple and governor's residence would be set up. This pattern bears a striking resemblance to the Egyptian city hieroglyph – the cross within a circle – and large and small versions of it proliferated from Algeria to Germany, from Syria to Britain. We have very few accounts of what life was like in a Roman city, but one testimony by Libanius has survived, describing Antioch around the year 300 AD, which has a remarkably modern ring:

'*As you walk along you find a succession of private houses, with public buildings distributed among them at intervals, here a temple, there a bath establishment, at such distances that they are handy for each quarter and in each case the entrance is in the colonnade ... it seems to me that the pleasantest and the most profitable side of city life is society and human intercourse, and that that is truly a city where these are most found. It is good to talk, and better to listen, and best of all to give advice, to sympathise with one friend's experiences, sharing their joys and sorrows, and getting like sympathy from them; these and countless other blessings come of a man's meeting his fellows. People in other cities who have no colonnades before their houses are kept apart by bad weather; nominally they live in the same town, but in fact they are as remote from each other as if they lived in different towns ... Whereas people in cities lose the habit of intimacy the further they live apart, with us the habit of friendship is matured by constant intercourse, and develops here as much as it diminishes there.*'

One feature of Antioch's streets of which Libanius was especially proud was its system of street lighting, by means of numerous burning lamps: 'We citizens of Antioch have shaken off the tyranny of sleep; here the lamp of the sun is succeeded by other lamps, surpassing the illumination of the Egyptians. With us night differs from day only in the kind of lighting. Trades go on as before; some ply their handicrafts, while others give themselves up to laughter and song.'

In Rome itself, civic life centred on the great public spaces – the forum, gardens, the baths and the circuses; but, as in Athens, the majority of the population lived in very humble surroundings. Rome invented the first high-rise tenements, the six-storey *insulae*, slum-dwellings that were dark, overcrowded, without sanitation, and so shoddily built that they often collapsed, killing their inhabitants. Rome was notorious for its filth and its outbreaks of plague, in spite of the marvellous aqueduct which brought fresh water from the hills, and the famous sewer, the *cloaca maxima*, which served only the wealthy citizens who paid for it. The real poison in the civic life of Rome, however, was the circus, where scenes of appalling violence were enacted daily, leading to the pathological decay of civilised life, in an uncanny

parallel to the rituals of blood sacrifice which were developing at this time in Central America. With its population reputedly around one million by the first century AD, Rome was the greatest centre of monumental architecture, urban wealth, consumerism, public administration and institutionalised savagery that the world had ever seen. As Rome decayed, morally and politically, its world dominance and its magnetic power were transferred from the physical to the spiritual realm, giving birth to the idea of the heavenly city. The Christian Church as a spiritual authority replaced Roman secular power, and claimed an equally universal sway. Now it was that the Judaic idea of the purified, restored city was transferred to the heavenly realm, and the powerful symbol of the new Jerusalem began to capture men's imagination.

No complete contemporary maps of any of the ancient cities – of Ur, Babylon, Memphis, Thebes, Athens, Alexandria or Constantinople – survive, and we have no absolutely clear evidence that they even existed. In the case of Rome, however, land surveying and city planning were essential parts of the imperial process, and it would be most surprising if Rome itself had remained unmapped. In fact a massive and detailed plan, known as *Forma Urbis Romae*, was drawn up and incised into stone tablets around the year 210 AD. Parts of it survive in fragments, showing that it was drawn on the very large scale of 1:240, about twenty feet to one inch, so that it measured, in its entirety, no less than thirteen metres by eighteen. This scale was not entirely uniform, however, as there was a tendency to make the important public areas larger than the common residential streets. It showed the administrative divisions of the city, but we have no clear idea how it was actually used by the city authorities, or whether it could be updated. The *Forma Urbis Romae* is an exceptional survival, and nothing else on this scale has been found for any other city from the ancient or classical world.

o o o

In the violent movement of European peoples that occurred between 400 and 700 AD, and which we call the Dark Ages, urban life was dislocated and almost destroyed. The population of towns and cities shrank as the inhabitants fled the barbarian invaders: in those days an inconspicuous hideout in the countryside was better than a palace in the town. Rome itself became depopulated, grass-grown and ruined, a fate shared by cities throughout the former Empire. But this decline of the city was counterbalanced by the rise of a new type of community: the Christian monastery emerged as effectively a new *polis*, which nurtured a new form of the common life. This life was of an unusual kind, being based on religious values, which appeared to deny the conventional bases of secular social life – the accumulation of power, prestige and wealth. Yet from this seed, a new phase of urban development would eventually arise in Europe, for it was the monasteries which contained and transmitted both the culture of antiquity and the ideal of the common life. Moreover, in

The Jerusalem of the crusaders, depicted as a medieval European city.
The British Library, Royal MS 1.e.IX, f.222

the social chaos of the Dark Ages, the Christian Church acted as Europe's unifying force, so that abbots and bishops acquired temporal powers, because no other responsible authority existed.

This bleak picture of post-Roman urban decline is true of western Europe only. A study of the world map around 700 AD would reveal that urban life was still flourishing in the Byzantine region, in the Islamic world, in India and in China. The Muslim powers in particular secured their realms, from Spain to Afghanistan, by conquering established cities like Damascus or Toledo, and by founding new ones, such as Baghdad and Cairo, just as the Romans had done before them.

By the ninth and tenth centuries, a recovery of urban life in Europe was under way, and it occurred in two phases, military and economic. The first phase sprang from the need for protection, especially from the threat from the Norse raiders, when even London and Paris were attacked and looted by fleets sailing up river from the coasts. City walls were rebuilt, and in a few cases (at Nimes and Arles for example) the old Roman amphitheatre was actually converted into a small, fortified town. The popula-

tion that had earlier deserted the towns now began to flow back, reviving the town's role as a container of food and of skills. Markets were held to feed the growing urban populace, selling food grown in the immediate locality, and more exotic goods brought from distant areas. Those who came to these markets were protected by the 'market peace', symbolised by the crosses erected in the market-places. In addition to the older royal and clerical strongholds in the city, newer merchant quarters appeared. The crucial process fostering this urban revival in Europe was the conversion of its pagan peoples to Christianity, bringing them under the influence of a single, recognised civilising authority: this created the more stable environment in which city-to-city trade could begin to flourish. Cities across Europe were linked by long-distance trade routes through the Alps, while port cities received and passed on goods from the Baltic and the Mediterranean, facilitated by the growth of credit and banking. Town and country interacted as the growth of urban populations stimulated the demand for food.

The years between 900 and 1300 saw the founding of thousands of new towns throughout Europe, and

The heavenly and the earthly cities, from Augustine's *City of God*; virtue reigns above, the seven deadly sins below. The disparity between the real and ideal city was a constant theme in medieval theology.
Bibliothèque Nationale, Paris

these became the agents of a fundamental shift in social and political life, encapsulated in the expressive German phrase *Stadtluft macht frei* – 'town air makes you free'. This refers to the practice whereby feudal lords released citizens from their obligation to feudal service in exchange for rent money. The feudal lords were willing to foster the growth of towns within their domains because they offered them a cash income. This urban freedom was embodied in the town charter, which enshrined the liberties of the citizens in documents which were signalling the coming end of feudalism. The citizens were made responsible for their own defence too, which made the charter doubly attractive to the overlord. Other urban sources of income – customs, tolls and fines of various kinds – were often shared between the town and the lord. Thus the medieval town was defined not by any physical structure, but by the existence of groups of free citizens working side by side. The citizens would band together into guilds or other

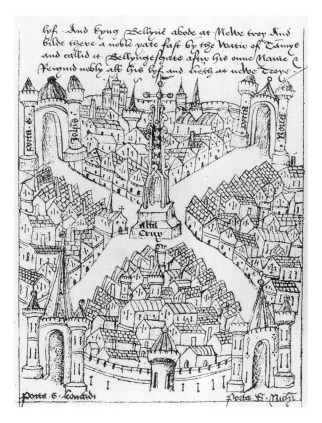

hyf And kyng Bellyn abode at Nehe troy And
blde there a noble pate faft by the Watic of Tamye
and called it Bellynggate after his onne Name
Reynid nobly aft his lyf and lieth at nehe troye

The archetypal medieval town: four gates at the
cardinal points in the town walls lead to the market
cross; Bristol in 1479.
The British Library, Ac 8113/99

§ GENVA §

Genoa from the *Nuremberg Chronicle*, the first
of a new generation of city-views: lively, evocative,
imaginative rather than accurate, they helped to foster
the Renaissance taste for city pictures.
The British Library, IB 6422

associations for mutual support, for in the Middle
Ages the unattached individual was nothing – he was
excommunicated or outlawed – and from these guilds
he derived his identity. Town life was deeply bound
up with these associations, and by its shared religion,
while the most characteristically urban institution
was the university, which was really a guild of scholars dedicated to preserving and disseminating classical and religious learning. Religious observance
acted as a unifying force in medieval urban life,
evoked in this description, penned by Albrecht
Dürer, of a great procession in Antwerp in the 1490s:

'On Sunday I saw the great procession from the Church of
Our Lady at Antwerp, when the whole town of every craft
and rank was assembled, each dressed in his best according
to his rank. And all ranks and guilds had their signs by
which they might be known. In the intervals great costly
pole-candles were borne, and three long old Flemish trum-
pets of silver. There were also, in the German fashion,
many pipers and drummers. All the instruments were
loudly blown or beaten. I saw the procession pass along the
street, the people being arranged in rows, each man some
distance from his neighbour, but the rows close behind the
other. There were the goldsmiths, the painters, the masons,
the broderers, the sculptors, the joiners, the carpenters, the
sailors, the fishermen, the butchers, the leatherers, the cloth-
makers, the bakers, the tailors, the cordwainers – indeed
workmen of all kinds, and many craftsmen and dealers
who work for their livelihood. Likewise the shopkeepers and
merchants and their assistants of all kinds were there. After
these came the shooters, with guns, bows and crossbows,
and the horsemen and the foot-soldiers also. Then followed
the watch of the lord magistrates. Then came a fine troop all
in red, nobly and splendidly clad. Before them however went
all the religious orders and members of the foundations,
very devoutly in their different robes. Twenty persons bore
the image of the Virgin Mary with the Lord Jesus, adorned
in the costliest manner, to the honour and glory of God.'

The medieval city is being presented here as the
uniquely apt stage for the tableau of life's functions,
secular and religious: the city calls forth the myriad
human talents, and all citizens had their part to play,
from the richest to the humblest.

It is from this period – the late Middle Ages – that
we have our first real images of Europe's cities,
sometimes as tiny pictures in manuscripts, sometimes as backgrounds in paintings, and sometimes
as woodcuts in early printed books. The city that
was found depicted most often throughout the Middle Ages was Jerusalem, whether as a stylised clus-
ter of towers, or in a more genuine attempt to evoke
the individuality of the city. From all of these we
receive the characteristic impression of church
spires rising above narrow, winding streets, turreted
walls pierced by heavy gates, or frequently a harbour or river-frontage packed with small craft, the
very impression which we still receive today when
visiting Rothenburg, Durham, Carcassone or La
Rochelle. The most obvious difference in scale
between these towns and anything that came later
was the narrowness of the streets, for virtually all
movement was on foot and wheeled transport

almost completely absent. Chief among these early visual records is the great *Nuremberg Chronicle* of 1493, which included vigorous woodcut views of more than one hundred leading cities, most of them quite stylised and inaccurate, but some of them showing authentic elements that must have been based on first-hand knowledge. Early printed books of travel also included images of cities, such as Breydenbach's *Pilgrimage to the Holy Land* of 1486, which included a particularly impressive view of Jerusalem, while manuscript sea-charts often included recognisable views of ports, as in Buondelmonte's *Book of Islands*, which was copied and re-copied in the fifteenth century. The 1480s also saw the publication of the first broadsheet view of a town, a very fine engraving of Florence by Francesco Rosselli, issued as a print in its own right, not as an illustration in a book.

The obvious feature of all these early town images is that they are not maps or plans in the accepted sense, but pictures, bird's-eye views or panoramas. They have often been created with considerable artistic skill and offer an imaginative view of a city

The supreme example of this new urban art would appear in the year 1500, in the breathtaking aerial view of Venice by Jacopo de Barbari, which evokes the unique splendour of the maritime capital of the Mediterranean. This new style of artistic city-image was undoubtedly a direct reflection of a new phase of urban history which began at this time. This kind of city-panorama was to prevail for fully two centuries more, before it was superseded by the city-plan in the more modern sense.

This new feeling for the city's aesthetic magnificence was clearly an aspect of Renaissance thought. Architects such as Alberti and Brunelleschi began to think of the city as a whole as a subject on which to exercise their intellect and their imagination, while their patrons – the Italian city rulers and merchants – were easily persuaded to open up the medieval warrens and replace them with noble streets, houses, squares, statues and fountains. The Renaissance study of perspective led to a concern with the vista, with the harmony that should exist between different buildings and, indeed, with theoretical plans for the ideal city – not ideal in the older philosophical

Amsterdam in the seventeenth century. Waterfront views, with craft large and small thronging the river or the harbour, became an immensely popular art form: the ships seem to symbolise the city's links with the outside world, with the sources of its wealth, and its political importance.
www.bridgeman.co.uk

from a vantage point which the human eye could never actually attain. Their purpose is clearly to present to us, directly and vividly, a picture of the living city, emphasising principal buildings, its walls, its surrounding hills or its harbour, and often showing its citizens too. They were not functional documents and could scarcely have been used for purposes of city administration. Their spirit is aesthetic rather than cartographic, and they seem to spring from a new-found pride in the majesty of these fine cities.

sense of law and government, but in the aesthetic sense; of course it was believed that the one was intimately linked to the other, that good government meant civic peace and civic beauty. The essential framework was ordered space, both in the physical and social sense, within which individuality of design and of personality could flourish. This was the rebirth of city planning in Europe, for the first time since the Roman era.

o o o

ARGENTINA. Straßburg.

1. Die Rhein bruck
2. S. Claus in Vndis
3. New thor
4. S. Wilhelm
5. S. Steffan
6. S. Catharina
7. Gulden thurn
8. Zum Rewern
9. S. Andreas
10. Münster
11. Zeughauß
12. Die Pfaltz
13. Predyger Closter
14. Jung S. Peter
15. Pfenning thurn
16. Barfüsser Closter
17. S. Niclaus
18. Spitel thor
19. Spital
20. Aller Heiligen
21. Frawen bruder
22. S. Thoman
23. Alt S. Peter
24. S. Marcur
25. Im Brüch
26. Steinstraßer thor
27. S. Johann
28. Heilig Grab
29. Augustiner Closter
30. S. Michael
31. S. Margretha
32. S. Aurelia
33. Deutsch Hauß
34. Weiß thurn
35. Cronenburger thor
36. Juden thor
37. S. Clara weerth
38. Eßscher thor
39. Schieß rein
40. Spital Meul

Left: Strasbourg in 1657, showing the massive bastions thrown up around so many European cities in the war-torn years of the seventeenth century.
The British Library, Maps C.25.b.19

These Renaissance ideals developed in the north Italian cities in the late fifteenth century, but they were merely the prelude to a second and more grandiose phase of deliberate city planning. The baroque city of the sixteenth and seventeenth centuries emerged as the expression of political absolutism, when the city became the setting for displays of royal power and self-glorification. Ironically perhaps, this movement really began in Rome, where a succession of Popes was determined to rebuild the tangled, half-ruined city in a glorious manner suited to the capital of Christendom. In the 1580s, Sixtus V commissioned the architect Domenico Fontana to drive wide, smooth streets through the ancient ruins, in order to link the principal pilgrimage churches. Piazzas were placed at focal points that gave panoramic views across the city so that, in Fontana's words 'beyond the religious purposes, these beauties provide a paradise for the bodily senses'.

This radical remodelling of a city would be widely imitated by Europe's rulers as, to a greater or lesser extent, they reshaped their capitals: Paris, Vienna, Berlin, Potsdam, Madrid and St Petersburg became symbols of royal glory, and it is noticeable that a number of the architects employed to oversee this process, such as Bernini, were also theatre designers. In one case, that of Karlsruhe, the new city was planned as a vast wheel whose avenues all radiated from the royal palace, the symbolic centre of the world. The idea of the city as a deliberately planned royal showcase spread to Persia, to the Isfahan of

Shah Abbas I. The Portuguese reproduced the magnificence of Lisbon in Goa, while in the Americas the energy and magnificence of the baroque were used in Mexico, Cuzco, Santo Domingo and Cartagena and other cities, as suitable symbols of imperial power in an alien land. The court at the city's heart was counterbalanced by a new feature at its edge: massive new fortified walls consisting of star-shaped bastions surrounded most of Europe's cities, as defences against increasingly powerful cannon, and increasingly aggressive monarchs whose natural political instrument was war.

This age of urban planning and urban magnificence found its perfect expression in the books of city views which now began to appear from the presses of the great European map-makers, beginning with Braun and Hogenberg's *Civitates Orbis Terrarum*, published in six volumes between 1572 and 1617. These splendid views of no fewer than 546 of the cities of Europe, the Middle East, Asia and the Americas were drawn from the best source materials available, and became an enduring success, adding a new dimension to the concept of armchair travelling. Robert Burton wrote in his *Anatomy of Melancholy* in 1621: 'To some kind of men it is an extraordinary delight to study, to look upon, a geographical map, and to behold, as it were, all the remote provinces, towns, cities of the world ... What greater pleasure can there be than to peruse those books of cities put out by Braunus and Hogenbergus ... ' All the Braun and Hogenberg maps were drawn on the

Opposite: A section of Barbari's bird's-eye view of Venice, 1500. A supreme product of Renaissance art, depicting the city from above, from a god-like viewpoint, it appealed powerfully to the imagination and to civic pride, and was hugely influential.
British Museum, 1895-1-22-1192

perspective-view principle, following Braun's own ordinance that 'the towns should be drawn in such a manner that the viewer can *look into* all the roads and streets, and see also all the buildings and open spaces'. Such a perspective, or bird's-eye view, was able to reveal churches, castles, markets, walls, the

The city as a military target: officer cadets study siege theory, from a Prussian military handbook of 1726.
The British Library, 8825.h.35

London after the Great Fire: a detail from Ogilby and Morgan's map of 1676, the first such scaled plan of London.
The British Library, Crace II.61(19)

river (if the city were built on one) and something of the surrounding countryside, while in the foreground there was almost always placed a group of figures in local costume, perhaps intended to link the viewer to the life of each particular city. Followers of the model created by Braun and Hogenberg included the great Dutch map-making house of Blaeu, and the German Matthaus Merian, author of a panoramic twenty-one-volume series of topographical views entitled *Theatrum Europaeum*, published

from 1640 onwards. Delightful and historically invaluable as they are, neither the Braun and Hogenbergs nor the Merian views can be called maps in the strict sense. The attraction of the city view was further increased by the market for battle-plans and siege-plans from the incessant wars which engulfed so many of the cities of continental Europe.

All the varied features of the baroque city – its avenues, its palaces, its bastions – were evidence of a centralised plan and a centralised power: they seemed to show that the city could and should function like a mechanism, not like a spontaneous meeting place of men and ideas. This was entirely appropriate in the age of the new science and new materialism, and it was Descartes, the great representative thinker of the age, who penned a highly logical argument for central urban planning:

'It is observable that the buildings which a single architect has planned and executed are generally more elegant and commodious than those which several have attempted. Thus also those ancient cities which, from being at first only villages have become, in the course of time, large towns, are usually but ill laid out compared with the regularly constructed towns which a professional architect has freely planned on an open plain; so that, although the several buildings of the former may often equal or surpass in beauty those of the latter, yet when one observes their indiscriminate juxtaposition, and the consequent crookedness and irregularity of the streets, one is disposed to think that chance alone, rather than human will guided by reason, must have led to such an arrangement.'

The impressive splendour of the baroque city can still be felt in its long-term legacy: the universal feeling that this was precisely the correct form of architecture and city-plan to give expression to political power. Washington is sometimes referred to as the last baroque city, with its vistas and its monumental architecture, but in fact the nineteenth century persisted in employing the baroque style for public and governmental buildings, while New Delhi and the unrealised Burnham plan for Chicago of 1909, and indeed Albert Speer's dreams for Nazi Berlin in the mid 1930s, all show the enduring link between baroque symbolism and political power.

The court influence in royal cities in the seventeenth and eighteenth centuries led to a new phase in social history and in the history of manners. In the first place the new, wide, paved streets now became the ideal setting for wheeled traffic. This shift away from foot traffic through intimate, narrow streets was to become a permanent feature, with incalculable effects for future city planning, and it also visibly and instantly marked the difference between rich and poor. This separation became increasingly embodied in the class differences between districts of the growing city: in London, Paris and elsewhere, the large estates on the periphery were sub-let and developed into residential streets and squares where the growing army of courtly and administrative hangers-on could dwell in some style, no longer intermixed with artisans and labourers, as they had been for centuries. This process began to turn the city into a commercial venture, where it became possible to live by buying, selling and renting, and where streets and houses became containers of wealth and badges of status. This same group of people was served by a variety of new institutions and buildings which began to enrich city life: the hotel, the theatre, the pleasure-garden, the art gallery and the shop, in all of which gradation by wealth and class was central from the first. This form of social life proved so attractive that a number of resort towns grew up – Bath is the classic example – where the atmosphere was more relaxed, detached from the presence of the court, and whose sole purpose was to act as a stage for pleasure-seeking and social display.

The evolution of the baroque city, in establishing the concept of town planning, also established the town-plan as we now know it. Design innovations like those achieved in Rome, Paris, St Petersburg, Karlsruhe, Philadelphia, New Orleans or in London after the Great Fire could not be executed without accurate, scaled plans of a kind fundamentally different from the artistic view or panorama. The view as seen by the human eye had to be replaced by the mathematical, scaled plan, whose effect is to place the eye directly above each and every sector of the map, and to keep each part of the map in correct proportion with all other parts. Only in this way can a scientific, functional diagram emerge. This principle had been slowly emerging in the field of topographic maps between the years 1500 and 1700, and now the town surveyor too had to resist his natural delight in panoramic views and in picturing the features of a city in elevation, and restrict himself to a more austere and functional street-plan. One of the earliest examples is the landmark plan of the rebuilt city of London after the Great Fire, published by John Ogilby and William Morgan in 1676. The lasting appeal of the perspective-view can be seen in the superb Turgot plan of Paris of 1739, which only succeeds, however, because the draughtsman has cleverly widened the spaces between the streets. As late as 1753 the plan of St Petersburg published by the Imperial Academy of Sciences still showed selected buildings in elevation. The emergence of the town-plan did not of course kill the city-view, but it had the effect of separating the two. In the eighteenth century many artists, the best-known being the brothers Samuel and Nathaniel Buck, developed a new genre of topographic views, which were constructed always from a natural viewpoint, usually a convenient hillside. Highly detailed and artistic, the Buck views made no attempt to 'see into' a city's streets through an artificial aerial perspective in the way that the Braun and Hogenberg maps had done.

The North American city was, for the first two hundred years of its development, simply the European city transplanted to the new world, and reflected all the diverse types of urban life and styles of architecture. Quebec was the walled, hilltop city, Boston the harbour city whose layout had grown naturally, Philadelphia was the planned, geometric city designed in London, Washington the baroque capital, consciously emulating European models. Only later,

under new historic pressures, did specifically American cities spring up: San Francisco mushrooming as a gold-rush town, Salt Lake City as the home of a Utopian sect, Chicago as a strategic rail centre, New York as the principal port of entry from Europe and gateway for millions of immigrants. As these and

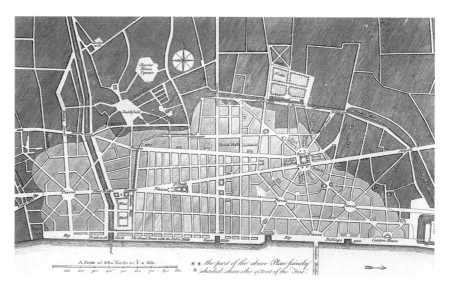

The London that Wren never saw: after the Great Fire, Wren proposed a magnificent new city, with avenues, vistas, piazzas, monuments and quaysides. Economy and expedience meant that it was never built.
The British Library, 576.m.15

The king as city planner: Peter the Great standing among the foundations of St Petersburg.
Ashmolean Museum, Oxford

other cities experienced dramatic growth, new building techniques led to the great discovery of modern urban design: that growth can be vertical as well as horizontal, and so the multi-storey skyscraper was born.

o o o

The ideal of the city as a grand design reigned supreme for three centuries, but by the year 1800 historic forces were at work in northern Europe which would undermine, indeed destroy, any possi-

Victorian London, a new phase of urban history: enterprise, machinery, wealth, squalor and seething humanity, portrayed in the art of Gustave Doré.
The British Library, 1788.b.20

Nineteenth-century Birmingham: magnificent civic architecture in the foreground, in the background the smoke of the working suburbs.
Birmingham Libraries

bility of city planning for many generations to come. The central process was industrialisation, but this was preceded by earlier seismic changes: in the late eighteenth century, and for reasons that have never been explained, Europe's population was increasing dramatically, while at the same time more efficient food production was reducing the number of people required to work the land. This already created the conditions for a shift of population to the cities, and to this was added the revolutionary new practice of factory working. In all manufacturing trades such as textiles and metal-working, the traditional place of work was the home or the small workshop. The arrival of usable steam power from the 1780s onwards made it far more efficient to bring workers together into factories, where they tended machines far more powerful and productive than any hand-processes could ever be. The factory was the core of a radically new phase of urban history. Beginning in England, but spreading rapidly throughout northern Europe and North America, the industrial city became the new urban form of the nineteenth century, some being old cities, including capitals, transformed by the growth of an industrial sector, others created anew around the factories. Formerly, cities had grown because they were centres of government, of religion, of defence or of trade; now they grew because they were centres of production. In Britain, France, Germany, Poland, Russia and America, the fabric of urban life was transformed by the explosion in population, by the huge, smoking factories in which they worked and by the endless rows of dark houses or apartments where they lived and died. The first phase of this process was concentrated wherever coal and iron were to be found, but soon the coming of the railway and then of electrical power created the potential for the industrialisation of almost any process in any location: the industrial city exploded across the surrounding countryside.

All this – the technical revolution and the urban growth – occurred without conscious planning, and some decades passed before the nature and the scale of these changes were noticed by social commentators, partly because they began in regions hidden from polite society. At first the appalling conditions in which the industrial population lived and worked were justified in terms of laissez-faire economics: toil, hunger and distress were necessary to restrict the growth of the poor, and their only escape must be through industry and self-discipline. Any attempt to alleviate their conditions would overturn the laws of self-interest by which the economy was believed to function. But by the 1840s the new mode of urban industrial existence had become a subject of concern for writers, philosophers, philanthropists and even some politicians. The novelists Elizabeth Gaskell and Charles Dickens, the philosophers Carlyle and Ruskin, and the reformers Chadwick and Shaftesbury all set out to show that this new urban environment was the creation of man, not of nature, and that it must be somehow be replanned and humanised. Of course the poor had always existed, and of course even the baroque capitals had contained scenes of squalor and misery – one has only to think of the London mirrored in Hogarth's paintings or in Fielding's novels, the London of the gin-shops, the brothels and the prisons. But what appalled the nineteenth-century critics was the dehumanising force of industrialisation, the vastness of the factories, the dirt and the smoke which they exhaled over a city, and the anonymity and hopelessness of the nominally free people who slaved there: this they rightly judged to be something new in human experience. George Gissing defined the city-dwellers as 'Men who toil without hope, and yet with the hunger of an unsatisfied desire'.

Out of the slums of the Victorian city, out of this toil and this desire, there grew many different creeds and many different blueprints for the future. At one extreme there was revolutionary Marxism – the conviction that in the industrial city, a historical process

City-dwellers had long been contemptuous of the countryside and its ignorance, yet the opposite impulse to escape the pressures of the city and return to the peace of the country has just as long a history, from Theocritus and Virgil onwards. Those who could afford to do so had always built houses just outside the city, where there was quiet, clean air and space for gardens. Such mansions lined London's Strand, west of the city, as early as the sixteenth century, while at the same time Chelsea became known as the 'village of palaces'. Other outlying villages such as Putney and Richmond also became fashionable settlements, full of elegant houses and extensive gardens. Such places were reserved at first for those with horses or private carriages, but in the mid nineteenth century they were democratised by the advent of the railways, and they became suburbs, within easy reach of the city centre. Where no such

was at work which must soon utterly reshape human society. At the other extreme was the rational, liberal pursuit of social planning as the solution to the city's ills. Labour laws, medical services, domestic architecture, water supply, sewers, schools, transport – all these things must be refashioned and integrated one with another, in order to transform the city into a civilised environment. To achieve these ends step by step within an existing industrial city was a daunting challenge, and hence there arose the alternative strategy of the new town, the garden city, or the model village, which should be built just outside the grasp of the industrial city, and where life could begin anew. But just as this solution began to be thoroughly explored, market forces and the people themselves produced their own solution – suburbia.

historic villages existed, the railways called new suburbs into being, mile after mile of tranquil middle-class streets, which enjoyed an era of idyllic peace for perhaps fifty years, between 1870 and 1920 – between the railway age and the automobile age. Such suburbs had at first a natural limitation, for they had to be accessible by foot from the railway station; but after 1920, the car removed this restriction, and the suburbs sprawled out of control. Suburbia was a movement to escape, a private solution to the public disorder and overcrowding of the city. The price exacted for this escape was the deadening uniformity of the later, car-age suburb.

The continuing spread of suburbs, even around smaller towns and cities, is testimony to the failure of the many planning solutions aimed at curing the

Chicago in 1892, a panorama of the fastest-growing city in the world, in the year when it hosted the Columbian Exhibition. www.bridgeman.co.uk

city's ills throughout the twentieth century. The most radical of these was the Corbusian city of high-density towers, which exercised an enormous influence on the practice of city planning the world over. The Corbusian design was intended to stop horizontal sprawl and to free space in the city for other uses. The result was grandiose towers in which nobody, given a choice, would live, and whose inhabitants reduced them to slums. Town planning is an inexact science, whose short-lived doctrines all too often create the problems of the next generation. Many of the Corbusian building projects of the 1950s and 1960s in European and North American cities were to be demolished before the end of the century. We cannot live without planning, but nor can we truly believe in the possibility of a planned city.

The true legacy of industrialisation was the commercialisation of the city, not only in the factory

An attempt to beautify the dynamic industrial city: Burnham's unrealised planned design for the Chicago Civic Centre, 1909, Art Institute of Chicago

itself, but in the office building. The office, a giant filing-cabinet full of human beings, is now the universal symbol of the modern city, as the cathedral was of the medieval city and the palace of the baroque city. The office, where millions arrive and depart each day, is the proof that it is work which is the driving force of today's city. They are complemented by the

shops which supply not simply food, but the essential trappings of our various lifestyles. The challenges facing the city are the housing, the transport and the security of its millions of inhabitants, but the challenges are always one step ahead of the answers, because of the pressures and changes which are born from commercialism. The commercial doctrines of newer, faster and bigger are constantly changing the city for motives of commercial gain, not for the rational or moral benefit of its citizens. The fundamental truth that emerges from a study of modern urbanism is this: the physical structure of a city arises from its functions, and today these are above all commercial functions. Against this simple fact, town planning, however well-intentioned, contends in vain. The city cannot be ordered, improved or humanised until the economic imperatives which drive it are contained and balanced by more humane forces. As long as commercialism determines the structure and function of the city, it will remain stressful, threatening, and dehumanised, because its structures reflect economic imperatives. Massive, virile and indestructible as it appears, the contemporary city and its people are now seen to be desperately vulnerable, both to weaknesses in its own workings and to outside enemies.

But, it will be argued, has not the city always been a centre of commerce? Were not medieval Cologne, Renaissance Venice, Amsterdam in the seventeenth century or Philadelphia in the eighteenth, centres of wealth derived from manufacture or trade? Undoubtedly they were, but the pressure of commerce was always counterbalanced by the recognition of other, civilising forces – religion, art, social pleasures and civic duties. Nor were these cities, or any prior to the year 1800, exposed to such pressures of growth, population, transport and public welfare as those thrown up in the industrial era. The power that the great cities of the world exercise over us is precisely that they embody the physical beauty and the intellectual legacy of the past. We imagine that they must contain or symbolise wisdom, or beauty, or a form of life, which we have forgotten or lost, qualities which are often embodied in the memories of the great figures who lived there. Why else do we flock to Athens or Rome or Amsterdam or Salzburg, except to see where Plato or Caesar or Rembrandt or Mozart walked, and to commune with the surroundings which inspired them? Like Henry James, we feel that here is history 'in the stones of the street and the atoms of the sunshine'. Of course this feeling must be largely an illusion, for great parts of all these cities have been rebuilt many times over, stone by stone, while the ancient streets are surrounded and overlaid by the torrents of modernity. Yet somehow their individuality remains, as the identity of the human being remains, despite the constant replacement of all the individual living cells. The city has functioned as a mirror to the history of civilisation, and it is sometimes possible to trace the evolution of the Roman garrison into the medieval fortress, the baroque showpiece, and then the industrial sprawl; but the process of attrition has been never-ending, and many of these stages are present only to our imagination.

The contrast between the monuments of the past and the challenges of modern urban living raises afresh the fundamental question 'What is a city?' Physically it may be mere stones and streets, but historically it was a concentration of energy, skill and wealth in the hands of a group of people who then sought and achieved a measure of self-government, the freedom to direct their own community. The central aim of city rulers was to regulate and enrich the social life and the physical fabric of their cities. But who were these rulers? Who actually created the great cities of the past? Who decided the layout of the walls, palaces, squares, dwellings and quaysides of Alexandria, Philadelphia, London, Beijing or St Petersburg? This is where political history and architectural history merge, and this diverse list suggests its own answer: cities have clearly been shaped and reshaped either by the individual will of an autocratic ruler, or by the commercial drives of its merchant class. Urban history has been marked by a constant tension between the movement towards freedom and self-government, and the power exercised by an overlord. Both agencies have fashioned beautiful cities, but the mysterious alchemy of the process has never been reducible to any formula, and now seems to be lost. In the modern age, the planning function has been inherited by democratic city governments - planning the physical fabric which should in turn mould the social life. The paradox is that democracy renders the competing claims of economics, leisure, transport, aesthetics and spirituality impossible to reconcile; democracy has yet to create a Florence, an Isfahan, a Salzburg or a Lhasa.

What of the future? Can the city be controlled, humanised and planned? Can we learn from the great cities of the past, or are they merely museums, their achievements irrelevant, and unrepeatable in the modern age? A few years ago it was fashionable to predict the growth of the 'megalopolis', the supercity, extending over thousands of square miles, linking previously discrete centres. The clearest example was thought be the north-eastern United States, where link development from Boston to Washington may embrace as many as fifty million people in one urban sprawl. However, the technology of the computer age has placed a huge question mark over the traditional city: given the communications revolution, why do millions of people still need to gather each day in order to work, when that work consists in messaging through electronic screens - the same screens which they all possess in their homes? Does not this facility point to a radical slimming down of human traffic in the city, and therefore a solution to the pressures on transport and services?

In fact it is obvious that commerce and administration are still clinging to the city, especially the metropolitan and capital city, and refusing to leave. The power and prestige of the city-centre location are still not felt to be outdated; but there is something deeper than that. Cities bring together cultures and ideas - this has been their historic role. They convert human power into form, energy into civilisation. They are like brains, directing and developing civilised life, and without their innovative force the body would atrophy and cease to change and develop, and this, in modern thought, is the equivalent of death. The arts, which flourish naturally in the city, have always been seen as symbolising this energy, this exposure to new ideas, and have always provided one of the great attractions of city life. As long as the two principles on which modern commercial life is based - change and innovation - reign supreme, even technology will not outdate the city. The container, the magnet, the meeting place will remain essential to life's processes. And perhaps,

despite all the city's vices and shortcomings, we should be glad of that: if a retreat into isolation were to be the end to which human civilisation had been leading, that would indeed be the most damning judgement on five thousand years of urban history.

The heavenly city come down to earth - a perpetual ideal, from a seventeenth-century engraving.
The British Library, 660.a.25

Above: The ruins of
Teotihuacan. This view
looks south from the
Pyramid of the Moon,
with the Pyramid of the
Sun to the left and the
two-mile-long Avenue of
the Dead leading away to
the right. From Karl E.
Meyer's *Teotihuacan*, 1973.
The British Library, x.425/1231

Right: Teotihuacan: the
ancient Mexican city
destroyed so mysteriously.
No contemporary map or
image of the city exists.
This highly accurate plan
of the main avenue and its
surroundings was drawn by
the American archaeologist
René Millon, and shows the
great ceremonial compound
and Temple of Quetzacoatl in
the south, the Pyramid of the
Sun centre right, and the
Pyramid of the Moon at the
northern end. From Karl E.
Meyer's *Teotihuacan*, 1973.
The British Library, x.425/1231

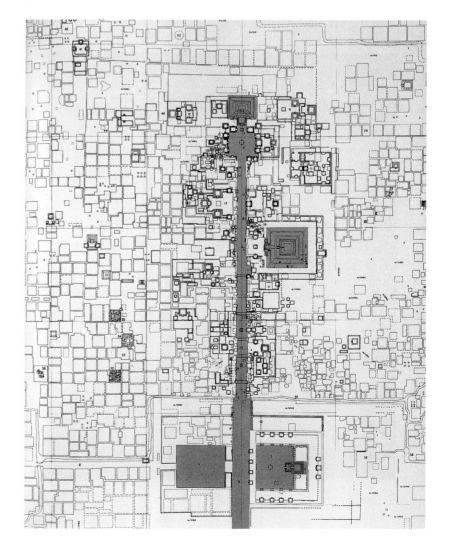

Teotihuacan

a vanished city

The city is man's grandest and at the same time man's most intimate creation and, as such, there is something uniquely evocative in the phrase 'vanished city' or 'lost city'. The idea of a towering, densely populated, vibrant city being abandoned, crumbling into ruins, to be lost under drifting sand, is profoundly disturbing. It overturns our sense of historical continuity, suggesting that civilisation may unravel, and time itself go into reverse. Yet it was a common enough event in the remote past. Some of the great vanished cities of the past have retained a clear identity because their names – Troy, Babylon or Nineveh – have been immortalised in literature, and their excavations by modern archaeologists became dramatic acts of rediscovery. But other cities – in some ways the most intriguing – have no such historical identity: their discovery has unveiled cultural epochs of which no one had previously dreamed. The origin, the way of life and the destruction of these cities remain shrouded in mystery, and few such mysteries are greater than that of Teotihuacan.

Until around 1900 the site of Teotihuacan was just a series of grass-grown mounds in the open country some 40 miles (65 km) north-west of Mexico City. Since that time, prolonged excavations have revealed what was undoubtedly the greatest city of the ancient Americas, the location of magnificent cultic buildings, and home to perhaps 200,000 people. Modern dating techniques have established that Teotihuacan was built around 200 AD, and it is clear that the city did not grow in a haphazard fashion, but was rigorously planned. The city's focus was a series of temples and pyramids which were disposed around a vast central avenue, known as the Avenue of the Dead. It was more than 2 miles (3 km) long and 130 feet (40 metres) broad, and punctuated by raised platforms, which suggest that the avenue was designed and used for processions and rituals. Towering over this avenue was the colossal Pyramid of the Sun, more than 200 feet (61 metres) high and 700 feet (213 metres) wide at its base, with a great stairway leading up from the avenue. The name Pyramid of the Sun was applied only in comparatively modern times, but at the summer solstice the sun's zenith is directly over the pyramid, causing its sides to be quite unshadowed. We can only conjecture that this moment was marked with a public ceremony, which may have symbolised the link between the city and the cosmos. On either side of the Avenue of the Dead are the remains of residential palaces, of great size and grandeur, where the elite of the city must have lived. Towards the outer edges of the site lay innumerable much smaller houses for the common people. The northern end of the avenue is marked by a second, smaller pyramid, the Pyramid of the Moon. Pyramids such as those at Teotihuacan have driven a number of historians to suggest that an historic link must exist between the civilisations of ancient America and that of Egypt.

What was Teotihuacan? Who were the people who built it, and what rituals were enacted there? Such a large city cannot have been self-sufficient, but must have been part of a long-range trade network which brought food and materials from the surrounding countryside. Was it planned as a ritual centre, as seems likely, and if so what were the beliefs which inspired its unique structures and layout? What was its social structure, and why have no clear indications been found of kingship or of military quarters? We know that the people of Teotihuacan were neither Mayan nor Aztec, but in the absence of any written records we have few clues to their identity or history. According to legend, the Aztecs visited the site centuries later, and named it 'The City of the Gods' because of the vast size and inexplicable nature of its ruins. The one secure fact is that around the year 750 AD, Teotihuacan suffered some catastrophe: it was ravaged by fire and abandoned by its people, never to be inhabited again. It may have been attacked from outside, but by whom is a mystery. It may have been destroyed by a popular uprising, perhaps in protest against a brutal theocracy and institutionalised human sacrifice. It may be significant that skeletons of bound individuals have been unearthed in their hundreds from beneath the temples. The city of Cholula, some 50 miles (80 km) to the south-east and site of a lower but still more massive pyramid, seems to have been abandoned at the same time, and it may have been a sister city.

At all events, the most elaborate city ever built in ancient America became almost overnight an empty ruin, leaving not a single memory behind and bequeathing to the future a series of insoluble mysteries. The stories of Teotihuacan and other vanished cities remind us starkly that history is not one straight line: they take us back to a time before global history, when civilisations were island communities, surrounded by destructive forces, or perhaps containing those destructive forces within themselves. In that context, a city or a state could be destroyed without trace, perhaps by outside enemies, or perhaps by an outraged populace. We cannot imagine such a catastrophic break with the past now, but over thousands of years it has been a fate which overtook many, many cities.

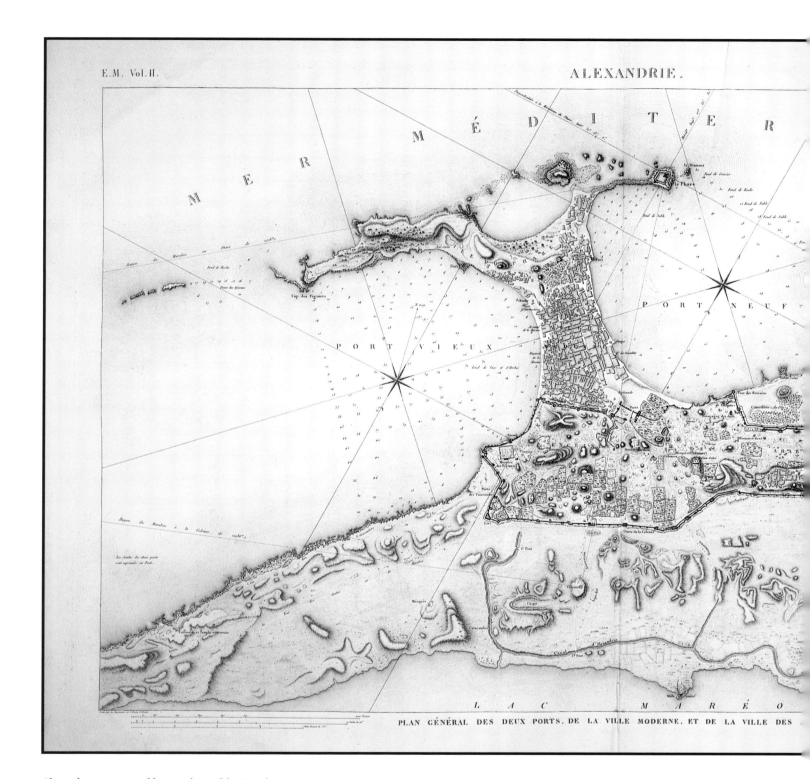

ALEXANDRIE.

MER MÉDITER

MER

PORT NEUF

PORT VIEUX

PLAN GÉNÉRAL DES DEUX PORTS, DE LA VILLE MODERNE, ET DE LA VILLE DES

L A C   M A R É O

Alexandria, as surveyed by members of the Napoleonic expedition in 1798, showing the city's unique setting on a slender arm of rock between sea and lake. In classical times Pharos was an island, but centuries of silting around the causeway have joined it to the mainland, creating twin harbours. The ancient city was immediately south of the 'Ville Moderne' named here on the peninsula, while the great Canopic Way ran directly east-west across the city.

The British Library, 1899.K.1 (74)

Alexandria

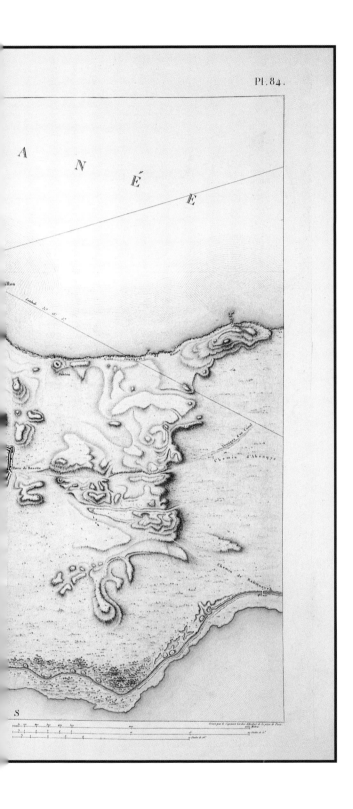

Pl. 84.

No ancient city has an origin so clearly recorded as that of Alexandria. It was founded in 332 BC by Alexander the Great to be his regional capital, and a naval base which would command the eastern Mediterranean. The site embraced the older town of Rakotis; it had an excellent anchorage at the small offshore island of Pharos; and a freshwater lake immediately to the south, fed at that time by a branch of the Nile. The architect was Dinocrates, commissioned personally by Alexander to build a city to rival any in the Hellenic world. The main thoroughfare, which followed the east–west axis of the city, was the Canopic Way, now called al-Hurriyah Avenue. This was intersected at its western end by the Street of the Soma, and at this intersection was built the celebrated Museum with its great library. Not a museum in the modern sense, but a research institute, dedicated to the muses of art and learning, it was staffed by scholars invited from the schools at Athens; its presence in the city was to make Alexandria the intellectual centre of the Hellenistic world. At the seaward end of the Street of the Soma were placed the two Egyptian obelisks later known as Cleopatra's Needles, which found their way eventually to London and New York. The most celebrated feature in Alexandria was actually in the sea - the great lighthouse raised on the island of Pharos, reportedly more than 400 feet (122 metres) tall, which became one of the seven wonders of the ancient world. A causeway led to the island, and this effectively created an east and a west harbour. A canal was dug to connect the city to the Nile, which in turn had been linked to the Red Sea by another canal. This was to prove vital to the city's prosperity, bringing Egyptian and foreign goods through the port. The Nile canal also supplied the city with water through a network of channels and cisterns, recently excavated.

Alexander never saw his new capital, as he departed for his military campaign in Persia, from which he never returned. On his death in 323 BC, control of Egypt passed to his viceroy, Ptolemy, founder of the royal dynasty. It was Ptolemy who brought back Alexander's body, and fashioned for it the magnificent tomb which was long to be seen in the city, but which has never been found by modern archaeologists. Various sites have been proposed and have excited the attention of scholars, but always in vain.

Egypt was a tempting target for Roman interest, encouraged by the military and personal weakness of the later Ptolemies in the first century BC. Events in Alexandria played a major part in the conflicts and intrigues which led to the imperial era in Roman history. Cleopatra, last of the Ptolemys, attempted to preserve her rule by captivating first Julius Caesar, then Mark Antony. Her defeat at the hands of Octavian at the Battle of Actium led to her death, and brought Egypt formally within the Roman Empire. It was in

One of the two great obelisks from Alexandria known as
Cleopatra's Needles, which were given in the nineteenth
century by the Egyptian government to the cities of London
and New York. They have no connection with Cleopatra,
but were dedicated at Heliopolis around 1500 BC and
re-erected in Alexandria in the Hellenistic era.
The British Library, 10094.h.l

the Roman period that Alexandria acted as a gateway through which Greek learning and philosophy were carried to Rome and the west: Euclid, Archimedes, Eratosthenes, Ptolemy and Plotinus all studied and taught in Alexandria. But the city was also a ferment of religious and occult ideas: neoplatonism, gnosticism, astrology, alchemy and the mystery religions all flourished here, and many of the earliest and seminal texts in these fields sprang from Alexandria. Philo-

Then in 1517 the Ottoman Turks conquered Egypt, and Alexandria became an unimportant outpost of their empire. The canal link to the Nile was permitted to silt up, as was the island anchorage of Pharos; for the next two centuries Alexandria contracted to the status of a fishing village. Its rebirth began in the early nineteenth century, with the revival of Egyptian nationalism which followed the Anglo–French campaigns of the Napoleonic era, the same campaigns which opened

sophical Judaism was represented by Philo, while the early Christian theologians, Clement and Origen, conceived their task to be the reconciliation of Christianity with the concepts of Greek philosophy.

As part of the Byzantine Empire, Alexandria was attacked first by a Persian force in 616 AD, and then in 642 it fell to the conquering Arabs as they progressed through North Africa towards Spain. The Arabs established a new Islamic capital at al-Fustat (Cairo) and Alexandria's long era of intellectual brilliance and political centrality was at an end. It survived as a major port, however, until the early sixteenth century, when the Europeans discovered the sea route to the east, bypassing both Alexandria and Constantinople.

European eyes to the culture and history of Egypt. The monuments of Alexandria's pagan age had been progressively destroyed by Christians and Muslims alike, and no ancient city displays so few traces of its classical past; today's city is entirely modern. Yet recent discoveries have been little short of spectacular, revealing beneath the city streets tombs of the fourth century BC - of the very first generation of Alexandrians. In the harbour, the underwater remains of a palace - perhaps Cleopatra's - have been found a short way out from the present shoreline. The greatest prize of all continues to fascinate the archaeologists: if the tomb of Alexander were to be finally located, it would indeed symbolise the recovery of Alexandria's brilliant and long-hidden past.

Alexandria: the island of Pharos with its great lighthouse, linked to the mainland by a causeway. The lighthouse was one of the wonders of the ancient world, and was the subject of many imaginary pictures such as this.
Index/www.bridgeman.co.uk

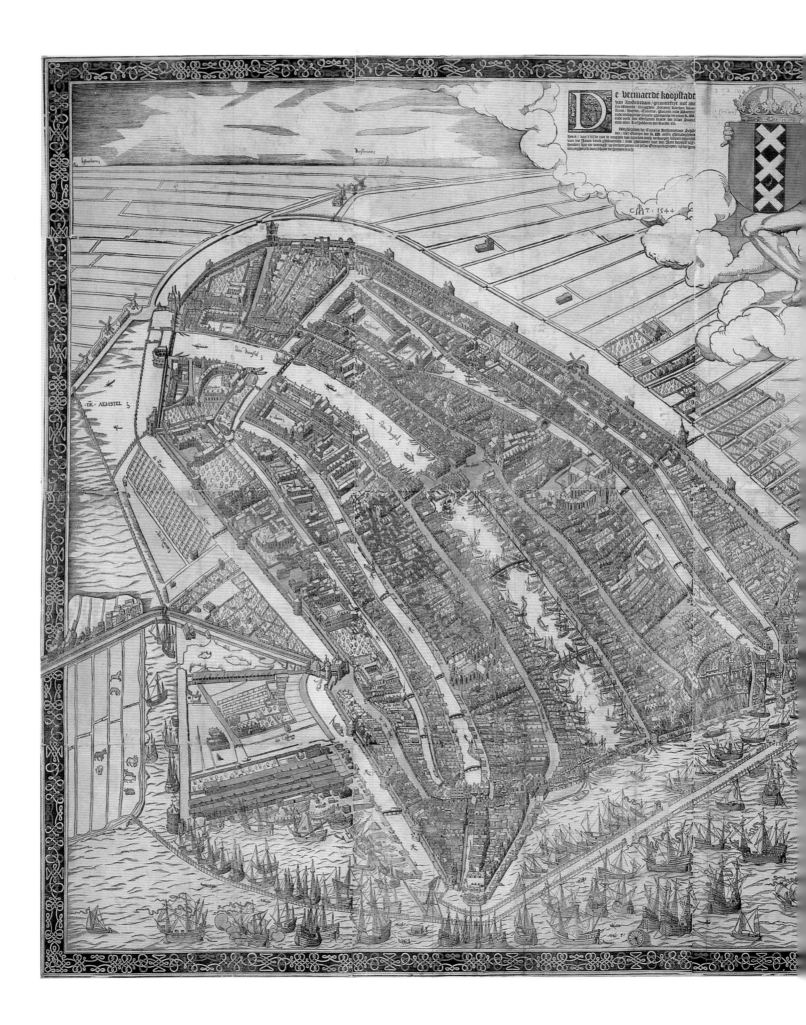

Amsterdam

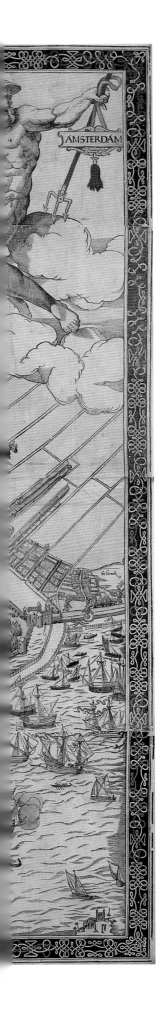

Amsterdam is a city of negatives, of understatements. It is not set against the grandeur of mountains, or on a superb natural harbour; it is not the capital of an empire, or even a seat of government; it is not a royal city, and has never been the scene of great historic events; it lacks majestic cathedrals, wide boulevards or imposing squares; it was never a magnet for artists or intellectuals – Rembrandt, its most famous son, was not a native – and it has not been immortalised by poets, novelists or painters. Amsterdam has in fact been a supremely successful bourgeois city. It was democratic, its architecture was picturesque, its citizens both tolerant and wealthy, and its urban texture was made delightful by the web of canals which intersect its streets.

The paradox of Amsterdam was noticed long ago, and it was expressed with typical rhetoric by Macaulay. 'On a desolate marsh,' he wrote, 'overhung by fogs and exhaling diseases, a marsh where there was neither wood nor stone, neither firm earth nor drinkable water, a marsh from which the ocean on one side and the Rhine on the other were kept out only with difficulty, was to be found the most prosperous community in Europe; the wealth which could be collected within five miles of the Stadhaus of Amsterdam would purchase the whole of Scotland.'

The canals and the sea-traffic which they carried lay at the heart of Amsterdam's history, for it was from the first a trading city. In the seventeenth century John Evelyn wrote, 'Nothing is more frequent than to see a whole navy of merchant ships environ'd with streets and houses, and every particular man's barque or vessel at anchor before his very door.' Originally a medieval fishing village, sited where a minor river, the Amstel, entered the Zuidersee, the river was soon embanked and a dam built to link the two halves of the community. In the later sixteenth century, the city became the focus of Dutch resistance to their Spanish monarch, especially with the fall of Antwerp to the Spaniards in 1585, an event which brought thousands of Protestant refugees to Amsterdam. At the same time, several thousand Jews, expelled by the Inquisition from Spain and Portugal, arrived to enrich the city's commercial life. Among these immigrants was the family of Spinoza, whose austere and reasoned philosophy seems to typify the spirit of Amsterdam. By the year 1600 the town on the Amstel had become the major European centre of shipping and banking. This role was further strengthened when the Dutch fleet – which had evolved from the North Sea fishing fleet into a formidable fighting navy – seized many of the Portuguese territories in the Far East, making Amsterdam the principal centre for European trade in spices, gems, coffee, silks, and other valuable cargoes. This was the material foundation for Amsterdam's golden age in the seventeenth century, when mercantile wealth was poured into domestic luxury in the form of architecture and works of art.

The medieval town remained for many years concentrated around the Amstel, enclosed within the Singel moat. With the dramatic increase in trade, further canals were dug – the Herencanal, the Keisercanal and the Prinsencanal. These brought goods and people to the doors of the merchants' houses, built at the very water's edge, with their distinctive gabled façades. Because of the marshy soil, huge timber foundations were first driven in beneath the new quays; it was said that whoever could see Amsterdam underground would see a huge winter forest. These three new canals formed successive concentric rings around the city centre, beyond which was built a fortified wall with watchtowers. This is the structure which can clearly be seen in this aerial view of 1544, which is remarkable for the hundreds of craft thronging the harbour or waiting in the Zuidersee, and for the gloating figure of Neptune, who symbolises the city's link with the sea.

In the nineteenth century, the city's prosperity declined as the large steamships were attracted to Rotterdam and Hamburg. The completion of the North Sea Canal in 1876 inaugurated a new era in Amsterdam's commercial history. Today, the aristocracy of wealth which shaped Amsterdam is less in evidence, but it still has a reputation for tolerance, permissiveness and radical thought. Like other cities built around water, it has a special atmosphere; but Amsterdam's is a relaxed, intimate charm, which is very different from the more obvious and much-praised beauty of Venice, Stockholm or St Petersburg.

Amsterdam, 1544, by Cornelis Anthoniszoon. Seen here from the north-east, this layout is still essentially that of the existing medieval town straddling the Amstel and enclosed by the Singel Canal. However, dwellings are beginning to appear to the west, where the new circuit of canals will be dug in the seventeenth century. The figure of Neptune may have been inspired by a similar image on Barbari's great panorama of Venice.
The British Library, Maps STA (4)

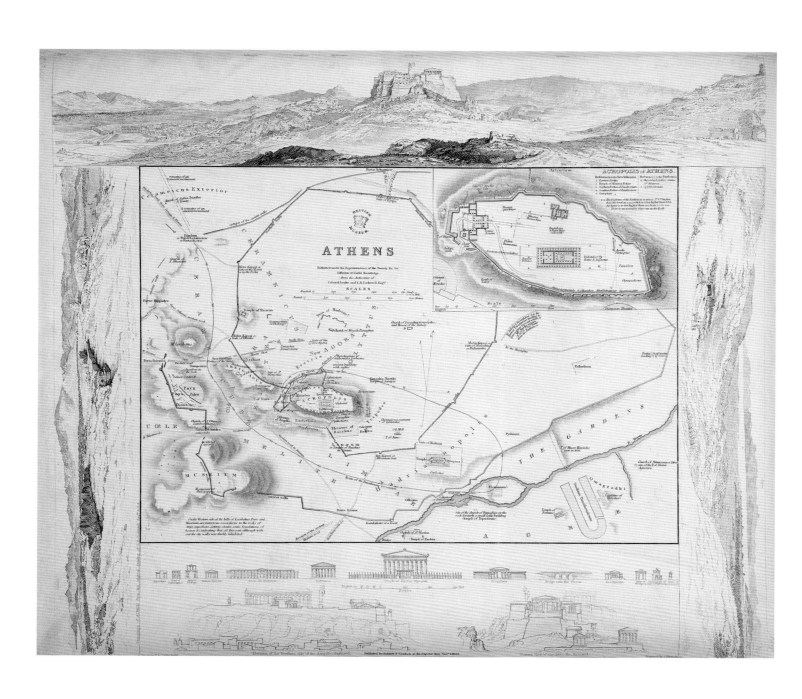

Athens

The city of Athens holds a unique place in the history of western civilisation, but its two centuries of intellectual brilliance were followed by almost two millennia of obscurity and domination by foreign powers. Of the ancient city only the magnificent fragments on the Acropolis remain, but our knowledge of it is exceptionally rich thanks to the many descriptions found in classical literature. The history of the city has really two distinct periods, separated by centuries of silence.

As early as 1500 BC, the hill of the Acropolis had been adopted as both citadel and sanctuary, and had been surrounded by a massive defensive wall. Settlement outside the Acropolis was confined at first to the south, with the area to the north used for burials. In the sixth century BC primitive shrines were replaced by impressive stone temples, and a broad ceremonial approach to the Acropolis was constructed, indicating that defence was no longer necessary here because the entire lower town had been walled; such a wall is described by Herodotus and others, but no trace of it has so far been found. Settlement spread to the north, and a new Agora was laid out – a broad open square for public events, which became the real heart of Athens. On one side of the Agora, a fountain-house received water from outside the city through a network of earthenware pipes. A theatre, council house, law courts and large private houses surrounded the Agora.

This already unique city was captured and burned by the Persians in 480 BC, before they were repulsed at the Battle of Salamis. Rebuilding commenced almost at once, but the Acropolis itself had to wait some thirty years before Pericles instituted the building programme which resulted in the Parthenon, the Propylaea and the Erechtheum. The new Acropolis not only consciously glorified the ideals of order and beauty, but also that of Athens's supremacy over the other cities of Greece. Like the ceremonial architecture of Egypt or Mesopotamia, it was designed to inspire and to overawe. The 'Long Walls' were built, linking the city with the port of Piraeus; these ran parallel for 4 miles (6.5 km), forming a corridor some 500 feet (150 metres) broad, allowing supplies to reach the city even in times of war. This was the city in which the philosophers flourished: Plato's Academy was a mile west of the Agora, while Aristotle's Lyceum lay to the east. The Stoics took their name from the *stoa*, the colonnade by the Agora, where they debated, while the Epicureans met in the gardens of their founder. It was the spirit of these schools – their dialectical pursuit of truth – which embodied the true legacy of Athens to the world, long after the city had fallen into ruins. Yet apart from the ceremonial centres, the city was reportedly mean and squalid. The streets were narrow and haphazard, their surface of dirt only; the houses presented a blank wall to the outside, being built around a courtyard; there were no sewers and no lighting; in some respects, much of Athens would have resembled a sprawling village. The population grew steadily, but following the principle that a city-state could only function if it were limited to a certain size, the excess population was continuously sent out to form new cities.

Athens was captured by the Romans in 86 BC, but it continued to prosper as an intellectual centre for several generations. New schools, temples, theatres and aqueducts were built, the Emperor Hadrian taking a special interest in embellishing the city, which he visited many times. While paganism survived, Athens was venerated as the home of philosophy, but the Christian Emperor Justinian closed the philosophical academies in 529 AD, an event often taken as symbolising the end of the classical era. Constantinople had already established itself as the capital of the Greek world, and Athens now gradually sank to the level of a small provincial town. Christian churches were founded, some in former temples, including the Parthenon itself. From the sixth to the thirteenth century, Athens's history is deeply obscure: it is almost never mentioned in documents of the period, and no archaeological remains had appeared to shed light into the darkness. Athens was occupied by the crusaders in 1204, the year when the crusading army also took Constantinople; the Parthenon became a Roman Catholic church and it was awarded a bell-tower.

In 1458 Athens fell to besieging Turks, who held the city for the next 375 years. Mosques were built, and an oriental element permeated the character of the city. The Parthenon suffered its greatest degree of structural damage as late as 1687 when the city was attacked by Venetian cannons. When Greece finally won its independence from Turkey and the newly invited King, Otto von Bayern, arrived in 1833, there was virtually no city, just a few thousand people living in a straggle of houses north of the Acropolis. The king was installed in the only two-storey stone house available, and began work at once in setting out the centre of the modern city, which spread west and north from Sintagma Square. Historians and archaeologists were commissioned to investigate and if possible to restore some of the ancient sites. Over the next fifty years, a programme of public building in the Greek-revival style gradually transformed Athens into a modern European capital.

Athens, 1844. No maps of classical Athens survive – if they ever existed. This map dates from just a few years after the emergence of modern Greece from centuries of Turkish rule. The new streets and dwellings of the Ottonian regime, north of the Acropolis, are not shown; instead western scholars seized their first opportunity for five centuries to investigate and map the city's classical sites.
The British Library, Maps 38.e.8

Athens: an early nineteenth-century view, when the Acropolis looked out across near-deserted countryside and a few unimportant dwellings.
The British Library, TAB 1237.a

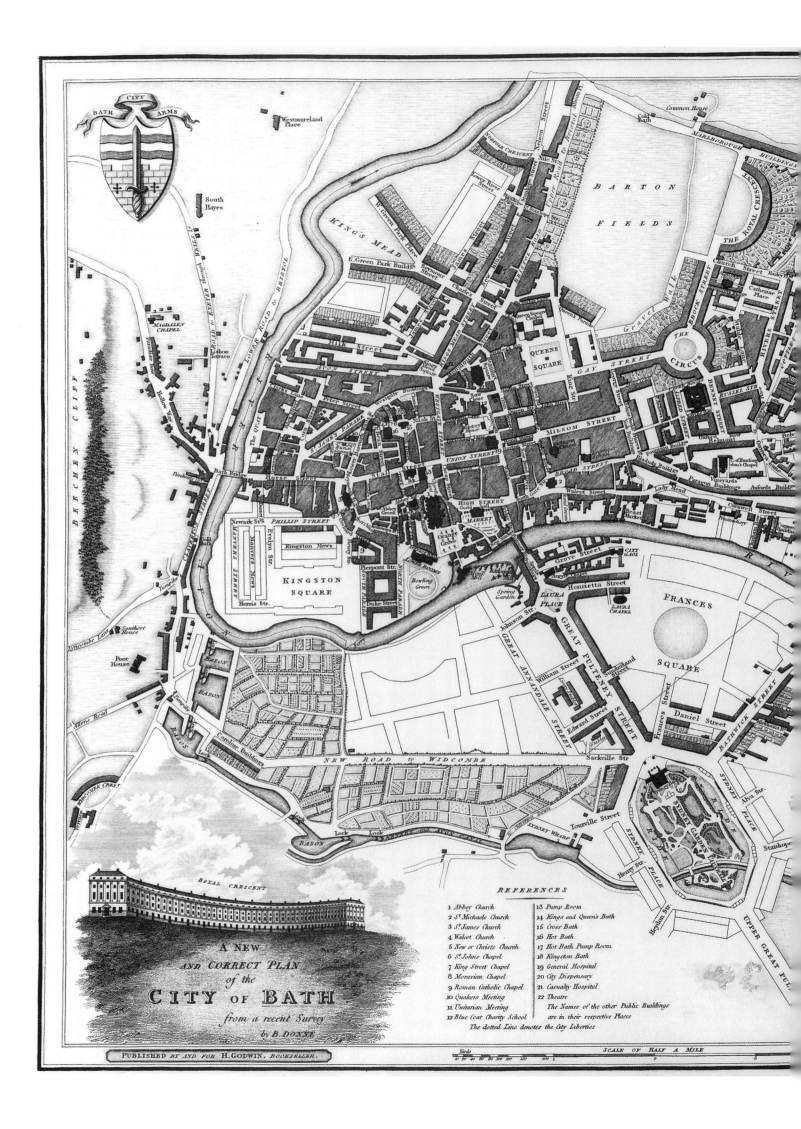

BATH CITY ARMS

Westmoreland Place

South Hayes

BARTON FIELDS

THE ROYAL CRESCENT

KING'S MEAD

Norfolk Crescent

Nile Str.

Common House
Cold Bath
MARLBOROUGH BUILDINGS

Lower River Street

Upper River Street

Nelson Street

Church Street

Catherine Place

BEECHEN CLIFF

MAGDALEN CHAPEL

Lisbon Terrace

ROAD TO EXETER through WELLS
LOWER ROAD to BRISTOL

E. Green Park Buildgs

Seymour Street

Charles Street

Milk Street

AVON STREET

Peter Street

The QUAY

Bath Bridge

Horse Street

St James's Parade

Nash Str.

BROCK STREET

QUEEN SQUARE

Queen Square Chapel

Queens Parade

GAY STREET

King Str.

THE CIRCUS

BENNET STREET

RUSSEL STR

RIVERS STREET

Milsom Street

UNION STREET

BROAD STREET

Walcot Street

Lady Mead

High Street
Guild Hall
MARKET

Orange Grove

Phillip Street

KINGSTON SQUARE

Kingston Mews

Pierpont Str.

North Parade

Duke Street

Bowling Green

Spring Garden

Grove Street

CITY GAOL

Henrietta Street

LAURA PLACE

LAURA CHAPEL

FRANCES SQUARE

Newark Str.

Evelyn Str.

Mayers Mews

Herris Str.

Southcot House

Avoncombe Lane

Poor House

Allene Road

BASON

BASON

BASON

Caroline Buildings

WIDCOMBE CRESCENT

BASON
Lock
Lock

NEW ROAD to WIDCOMBE

Johnson Str.

GREAT ANNANDALE STREET

William Street

Edward Street

GREAT PULTENEY STREET

Sunderland Street

Frances Street

Daniel Street

BATHWICK STREET

SIDNEY PLACE

RIDE

SYDNEY GARDEN

Alva Str.

Stanhope

Sackville Str.

Tourville Street

SYDNEY WHARF

Henry Str.

Heyton Str.

UPPER GREAT PUL

ROYAL CRESCENT

REFERENCES

1 Abbey Church
2 St Michaels Church
3 St James Church
4 Walcot Church
5 New or Christs Church
6 St Johns Chapel
7 King Street Chapel
8 Moravian Chapel
9 Roman Catholic Chapel
10 Quakers Meeting
11 Unitarian Meeting
12 Blue Coat Charity School

13 Pump Room
14 Kings and Queen's Bath
15 Cross Bath
16 Hot Bath
17 Hot Bath Pump Room
18 Kingston Bath
19 General Hospital
20 City Dispensary
21 Casualty Hospital
22 Theatre

The Names of the other Public Buildings
are in their respective Places

The dotted Line denotes the City Liberties

A NEW
AND CORRECT PLAN
of the
CITY OF BATH
from a recent Survey
by B. DONNE

PUBLISHED BY AND FOR H. GODWIN, BOOKSELLER.

Yards
10 20 40 60 80 100 120   150   200 1

SCALE OF HALF A MILE

Bath

The natural spring of thermal water on which Bath's fortunes have been built was doubtless known to the Celtic people of Britain, but it was the Romans who exploited it by building around it a complex of elaborate baths, and a town to serve them. The baths became a focus of healing as well as recreation, and were dedicated to the Celtic goddess Sulis and the Roman goddess Minerva; the Roman name for the place was Aquae Sulis. Virtually abandoned in the post-Roman era, medieval Bath became a centre of cloth manufacture, the seat of an Abbey and, briefly, a cathedral town. New baths were built and the waters' curative powers made the town both famous and fashionable. By 1600 physicians were recommending the virtues of the Bath waters, both to bathe in and to drink.

The city's great transformation – physical and social – came in the early eighteenth century, and it was consciously masterminded by a few individuals. Ralph Allen, entrepreneur and landowner, effectively commissioned the architect John Wood to rebuild the town centre from 1727 onwards in the neoclassical style, using graceful and durable Bath stone from quarries which Allen owned. Spacious squares, circles and crescents were laid out in what became a unique experiment in British town planning. Queen's Square, the Circus and above all the Royal Crescent were crowned with terraces that resembled Palladian palaces. Their location on rising ground above the town made their impact the more breath-taking. Beyond the river, Great Pulteney Street, 1,000 feet (300 metres) long and 100 feet (30 metres) wide, was designed by Robert Adam, and continued across the river via the unique bridge with its terraced shops. Assembly rooms, theatres and a redesigned baths completed the elegant new urban fabric.

The purpose of all this architectural energy was to attract and accommodate visitors, for Bath was becoming the earliest resort town in British social history. If medieval pilgrims visited shrines to improve their spiritual health, their eighteenth-century counterparts sought to refresh their bodies after years of gluttony and debauchery. Royal patronage began with Charles II and Queen

Bath, 1810, by Benjamin Donne. Drawn at the height of Bath's dominance of the world of fashion, this is the Bath Jane Austen knew. Rather strangely in a map of this date, west rather than north is at the top. The planned streets marked on the east of the river, such as Frances Square, would never be built, but would be laid out as parks. With the exception of the elegant lodgings of Great Pulteney Street, all Bath's attractions – assembly rooms and shops – lay west of the river.
The British Library, K Top XXXVII.23

Anne, and for the crowds of health-seekers who followed them, a rich web of social events and entertainments was spun: dining, music, dancing, theatre, gambling, promenading, shopping, all coordinated by a 'Master of Ceremonies', the celebrated Beau Nash, whose career was as central to Bath's brilliant image as Allen's or Wood's. Nash organised balls and assemblies, set rules for dress and conduct, and regulated the price of lodgings. But strangely perhaps, one of the great attractions of Bath was its egalitarianism: aristocracy, gentry, wealthy tradesmen, clergy and military all mingled in a way that was unthinkable in London. It was a high-class free-for-all whose appeal was immense. Scarcely a novel or play written between 1720 and 1820 fails to place a scene in Bath, or at least to mention it. Our image of the town is still largely conditioned by memories inspired by Smollett, Sheridan and Jane Austen. Wesley regarded Bath as 'Satan's headquarters', where the devil was virtually immune from attack.

Bath declined in popularity from around 1840 onwards. The steam age brought many other resorts within reach, especially those on continental Europe, while medical knowledge advanced beyond a faith in curative waters. Despite the ravages of modern development, much of Bath has survived, a unique monument to the Georgian civic ideal, where elegance of environment was regarded as the essential framework for elegance of living.

Bath, with its hillside setting and elegant architecture, was the subject of innumerable picturesque views such as this one.
The British Library, 199.i.7

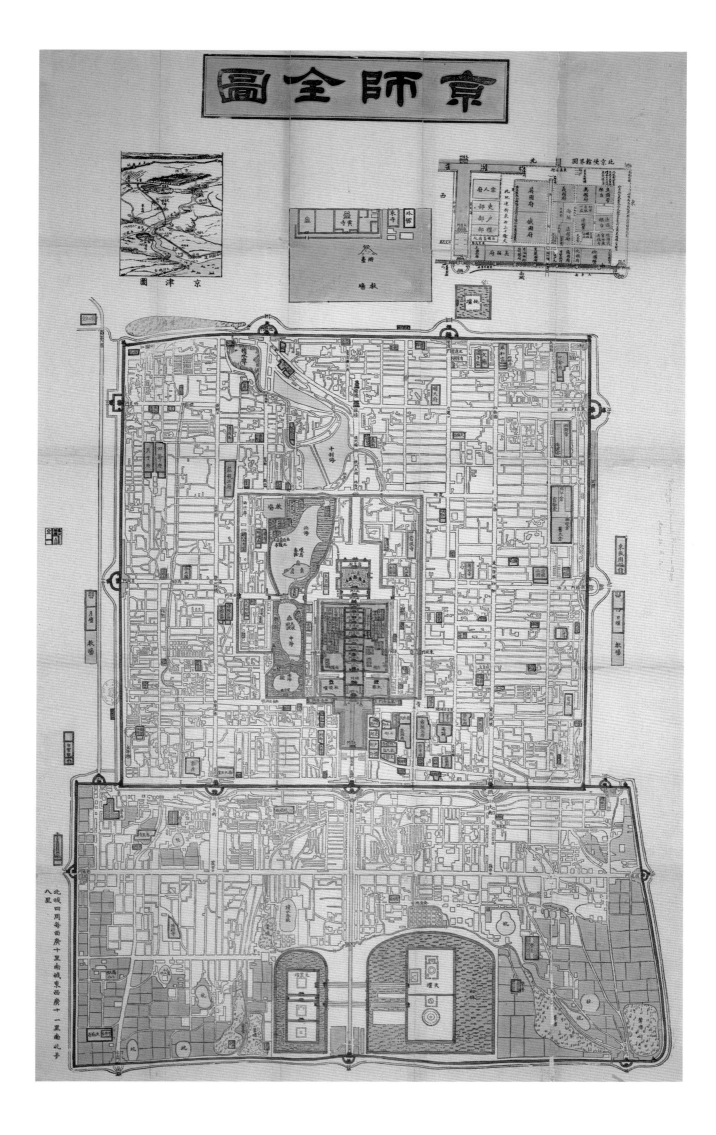

Beijing

Beijing originated more than 2,000 years ago as a garrison town near to the northern border of China. To the south lay fertile, well-populated lowlands, while to the north and west, beyond the Great Wall (which developed between the seventh and the third centuries BC), were mountains and the Mongolian plateau. By the twelfth century this city, then called Zhongdu, was the capital of the Jin dynasty which ruled northern China. It was from the north, from Mongolia, that Genghis Khan and his army streamed in 1215, breaching the Great Wall and taking Zhongdu, burning it to the ground. Half a century later, the great conqueror's grandson, Kublai Khan, rebuilt the city, gave it the Mongol name Khanbaliq, and made it the magnificent capital of Mongol China. This was the city which Marco Polo saw and described in the 1270s, with its palaces, temples, canals, bridges, and its population of perhaps half a million, greater than any European city of its day. The canals built at this time brought shipping from as far afield as the Yangtse into the heart of the city. None of the buildings of this period has survived, but the rectangular grid layout of the royal city is thought to have been inherited from Kublai Khan.

In 1368 Mongol rule was overthrown, and the establishment of the Ming dynasty signalled a new era in Beijing's history. Its current name, meaning simply 'Northern Capital', was bestowed by the Ming emperors, and they developed it as a series of walled cities, one within the other, with the royal enclosure known to the West as the Forbidden City at its heart. Orientated to the four cardinal points of the compass, the principle behind its complex of halls and courts was that no common eye should behold the Emperor and his family. Measuring half a mile from east to west and slightly more from north to south, historical records indicate that 200,000 men laboured for fourteen years in its construction. It housed also the huge bureaucracy which governed China. No part of it was overwhelmingly high, for its force came from the structures of seclusion which it embodied; the Hall of Supreme Harmony at 123 feet (37.5 metres) was for centuries the tallest building in Beijing. The strict rectangular form of the Ming city is easily visible in the modern city, for it was built around a central north–south axis. Walls, gates, temples, parks and markets were arranged symmetrically about this axis. Thoroughfares ran east–west and north–south across the royal city, linking its six gates and giving Beijing its chequerboard ground-plan, which Roman city engineers would surely have admired. The communist state has made no attempt to alter this plan, so that the centre of the modern city is still Tien an Men Square, where imperial proclamations were once given. As in many European cities, the course of the ancient walls - demolished only in 1949 - is now marked by the inner ring road and the metro line.

For five centuries the current of Beijing's history flowed on in total isolation from the outside world, until the European powers forced their way into China in the mid-nineteenth century. In 1859-60 British and French troops entered and looted Beijing, but left the Forbidden City untouched. Its sanctity was first violated by foreigners in 1900 after the collapse of the Boxer Rebellion and the flight of the royal family. Having miraculously survived decades of civil unrest during the twentieth century, the Forbidden City is now a meticulously preserved relic of the past in the heart of communist China. It was from a rostrum in Tien an Men Square that Mao Zedong proclaimed the birth of the People's Republic of China in October 1949, with Beijing as its capital. Rigid, centralised power has long characterised China, a power always prepared to use tyranny and brutality. This persistence of this sinister past is powerfully symbolised by the survival of the historic structures at the heart of modern Beijing.

Beijing: a Chinese woodcut map of c.1900, showing the strict chequerboard layout of the ancient city, which still survives. The Forbidden City is at its heart, with streets, gates and walls arranged around it in faultless symmetry. To the south stands the southern outer city, built in the sixteenth-century Ming period. As with many historic European cities, the lines of the walls are now marked by a ring road and a metro.
The British Library, Maps 30.b.54

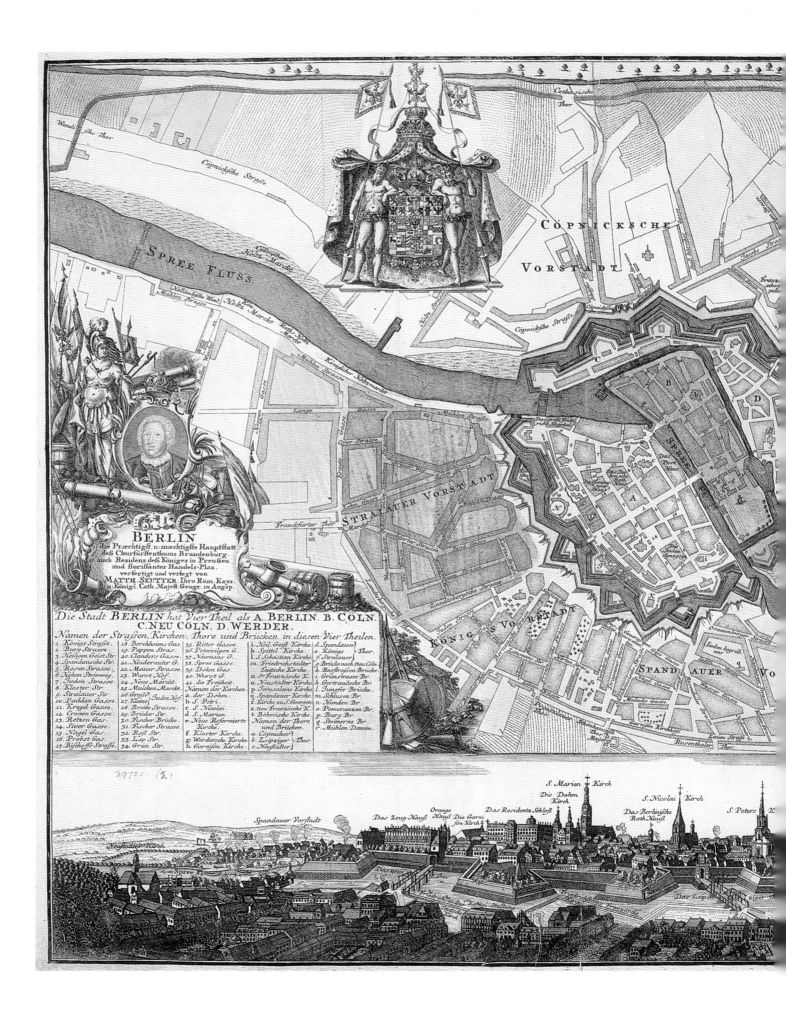

Wendische Thor

Cöpnicksche Strasse

SPREE FLUSS

CÖPNICKSCHE
VORSTADT

Cortbarische
Thor

Cöpnicksche Strasse

Cöpnicksche Strasse

Hollandische Wind Mühlen Strasse

Collnische Holtz-Marckt

Holtz-Marckt Rath-Nahe Marckt

Neuen Gasse

Lange Grune Gasse

Kleine Grune Gasse

Mühlen Strasse

Königlicher Holtzmarckt

Mühlen Strasse

STRALAUER VORSTADT

Franckfurter Thor

Marckt

KÖNIGS VORSTADT

SPANDAUER VO

BERLIN
die Præchtigst. u. mæchtigste Hauptstatt
deß Churfürstenthums Brandenburg.
auch Residenz deß Königes in Preußen
und florißanter Handels-Plaz.
verfertiget und verlegt von
MATTH. SEUTTER. Ihro Röm. Kays.
u. Königl. Cath. Majeſt. Geogr. in Augsp.

Die Stadt BERLIN hat Vier Theil als A. BERLIN. B. CÖLN.
C. NEU CÖLN. D. WERDER.

Namen der Straßen, Kirchen, Thore und Brücken in diesen Vier Theilen.

1. Königs Strasse.	18. Berchheims Gas.	35. Ritter Gasse.
2. Burg Strasse.	19. Pappen Strax.	36. Petersügen G.
3. Heiligen Geist Str.	20. Clanderer Gasse.	37. Naiman G.
4. Spandausche Str.	21. Heüderenter G.	38. Spree Gasse.
5. Rosen Strasse.	22. Mauer Strasse.	39. Dohm Gas.
6. Hohen Steinweg.	23. Wurst Hof.	40. Wurst G.
7. Juden Strasse.	24. Neue Marckt.	41. die Freyheit.
8. Kloster Str.	25. Mulcken Marckt.	Namen der Kirchen.
9. Stralauer Str.	26. Große Juden Hof.	a. der Dohm.
10. Padden Gasse.	27. Kleine	b. S. Petri.
11. Kregel Gasse.	28. Breite Strasse.	c. S. Nicolai.
12. Cronen Gasse.	29. Bruder Str.	d. S. Marien.
13. Retzen Gas.	30. Fischer Brücke.	e. Neue Reformierte
14. Siver Gasse.	31. Fischer Strasse.	Kirche.
15. Nägel Gas.	32. Roß Str.	f. Kloster Kirche.
16. Probst Gas.	33. Lap Str.	g. Werdersche Kirche.
17. Bischoff Strasse.	34. Grün Str.	h. Garnison Kirche.

i. Heil. Geist Kirche.	d. Spandauer
k. Spittel Kirche.	e. Königs } Thor.
l. S. Sebastian Kirche.	f. Stralauer
m. Friedrich-städter	g. Brücke nach Neu Cöln.
Teutsche Kirche.	h. Roßstraßen Brücke.
n. de Französche K.	i. Grün-straßen Br.
o. Neustadter Kirche.	k. Gertraudsche Br.
p. Jerusalems Kirche.	l. Jungfer Brücke.
q. Spandauer Kirche.	m. Schlusen Br.
r. Kirche zu S. Georgen.	n. Hinden Br.
s. neu Französche K.	o. Pomerannen Br.
t. Böhmische Kirche.	p. Burg Br.
Namen der Thore	q. Steinerne Br.
und Brücken.	r. Mühlen Damm.
a. Copnicker	
b. Leipziger } Thor.	
c. Neustadter	

29720. (2)

Spandauer Vorstadt

Neustadter Kirch

Spree

Das Zeug Hauß

Orange

Das Garni son Kirch

S. Marien Kirch

Die Dohm Kirch

Das Residentz Schloß

S. Nicolai Kirch

Das Berlinische Rath Hauß

S. Peters 2c.

Berlin

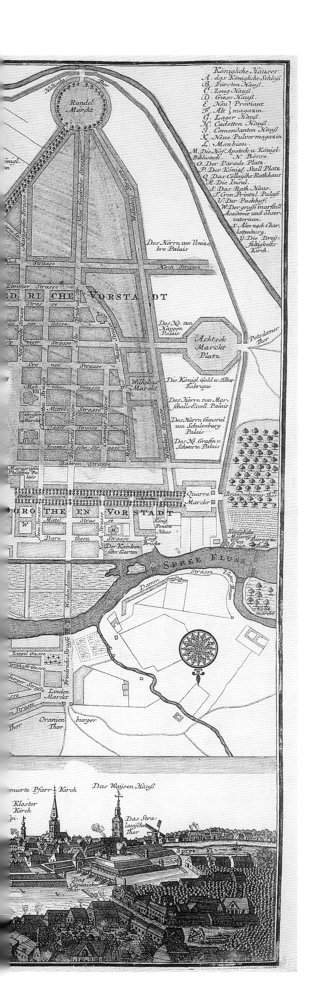

The tragedy of Germany during the most turbulent hundred years of her history is summed up in the fate of Berlin. The capital which was built on military power and was given a neoclassical façade, but whose rulers worshipped a mythology of blood and iron, was reduced to rubble, made subject to foreign powers, divided and embittered for two generations.

Berlin is unknown to history before the thirteenth century, when it was an unimportant settlement around an island in the River Spree. It was in fact one half of a twin-town pair: Cölln occupied the island itself, while a small fortified area on the right bank was Berlin, and the two would not formally merge until the late seventeenth century. Of the original medieval town, only the Marienkirche and Nikolaikirche now survive. Berlin's importance stemmed not from any natural advantages of geography, but from an accident of history, when the Hohenzollern dynasty became linked with the town. In 1411, Frederick of Hohenzollern, a member of a noble family from Swabia, far to the south, was appointed by the Emperor as Elector of Brandenburg, and Berlin became a *Residenzstadt*, a dynastic capital. The town remained small however, with a population of little more than 10,000 in 1650. The real history of Berlin begins only in 1640, with the reign of Frederick Wilhelm, known as the Great Elector. He restored the town after the turmoil of the Thirty Years War, fortifying it with modern bastions, opening canals from the River Spree, and incorporating neighbouring villages into one city. Even more important, he began the military and diplomatic process by which Prussia became the dominant power in the German-speaking world. It was his son who secured the title King of Prussia, and his great-grandson, Frederick the Great, who set out to make Berlin the equal of eighteenth-century Paris as a place of art and intellect, with Potsdam as its Versailles. It was in these years that Berlin's landmarks were created: Unter den Linden, the Tiergarten, the Charlottenburg, and the Gendarmenmarkt; the Brandenburg Gate as we know it was slightly later, built in 1789 as a more magnificent replacement for an old customs gate. Neoclassical architecture remained the favoured style for Berlin, combining Greek beauty with a certain Prussian rigidity. The architectural splendour was matched by an intellectual flowering. The court had attracted writers of the stature of Lessing and musicians like C.P.E. Bach, and when the university was founded by Alexander von Humboldt in 1810, its early scholars included Hegel, Fichte and the Brothers Grimm; among its later and less typical alumni were Marx and Engels. Nevertheless, the army still remained the basis of Prussian society and consciousness, and Berlin was always a garrison town, with tens of thousands of soldiers permanently in residence. Even the great Prussian army, however, was unable to resist Napoleon, who occupied and pillaged the city in 1806, sending back to Paris the bronze chariot which surmounted the Brandenburg Gate.

Berlin, by Seutter, c.1730. Both the plan and the view are turned to the south, perhaps to bring the imperial palaces and the churches into the foreground. The River Spree has been canalised to form a moat around the walls, intensifying the fortress-like impression. Yet the outer suburbs are already well developed, including Dorotheen Vorstadt, named after the Prussian queen. On the right, just above the river, the Brandenburg Gate is just visible. The cartouche surrounding the portrait of Friedrich Wilhelm I is suitably militaristic.
The British Library, Maps 29720 (2)

The Brandenburg Gate, symbol of Berlin's military past, completed in 1791. It was modelled closely on the Proplyaea, the gate of the Athenian Acropolis. The bronze chariot was briefly removed by Napoleon and taken to Paris as a victory trophy. Shattered during World War II, the Gate was rebuilt, but from 1961 until the reunification in 1989, it was shut off from both East and West Berliners.

In the early nineteenth century, Germany's headlong rush into industrialisation had a huge impact on Berlin, which lay in a region rich in raw materials and which now had good communications. From 200,000 in 1815, the city's population swelled to 1.2 million by 1880. The Great Elector's fortifications were dismantled in the 1860s and a *Ringbahn*, a circular railway, was laid in their place. Industries spread rapidly outside the Ring and a rash of poor, cheap housing appeared to accommodate its workers. Berlin participated in the 1848 street revolutions which swept Europe's cities, but the two hundred deaths failed to advance democracy in any real sense.

In 1871, thanks to Bismarck's diplomacy, Berlin achieved the goal towards which it had been moving since the reign of the Great Elector: Germany was united as one nation under the rule of the Prussian King, Wilhelm I, and Berlin became the capital of the Second Reich. These events were celebrated in the grandiose official architecture of the Victory Column and the Reichstag, the latter largely an empty symbol, for it wielded little real power. Berlin in the 1880s and 90s was paralysed in terms of internal politics, but it commanded the most dynamic economy in Europe, and its cultural life was enslaved by the military ideals of discipline and glory. To many, these three forces seemed to be propelling Germany towards some monumental crisis as the century drew to its close. The tensions and dangers of Berlin's social norms were beautifully captured in the novels of Theodore Fontane, where the rigid laws of Prussian society seemed, prophetically, to exact their human sacrifices.

The suffering of the population of Berlin during World War I from near-starvation was largely hidden from the outside world, as were the strikes and peace demonstrations which took place there from 1916 onwards. But in October and November of 1918, the edifice collapsed. A thousand people a day were dying from influenza and malnutrition; the armed forces mutinied; the Hohenzollern dynasty fled the city; the war was over; and a German Republic was declared on 9 November on the steps of the Reichstag. Street violence remained endemic however, and, deeming the city too dangerous for democratic politics, the government temporarily removed itself to Weimar. For a few brief years in the 1920s Berlin was famed as a place of avant-garde art and entertainment – the Berlin of the cabarets and the surrealist films. But the post-war malaise in German society ran very deep, and forces were already at work which were determined to reverse the catastrophe of 1918. Massive street violence heralded the Nazi rise to power, and continued after it, as the Nazis took control of social life. Hitler disliked Berlin and the Prussian establishment, and with his architect, Albert Speer, he nurtured grandiose dreams for the reshaping of the city on a monumental scale. War pre-empted these plans, and the Olympic stadium at Pichelsdorf is one of the few survivors of the Nazi era. Others may still haunt Berlin, however, as construction workers have recently exposed the Gestapo torture-cellars in the Prinz-Albrecht Strasse. These most shameful years in the city's history climaxed in 1938 on *Kristallnacht* (oddly on 9 November again) when synagogues and other Jewish buildings were destroyed, and the persecution became irreversible and deadly.

The Third Reich was to last a thousand years but, twelve brutal years after its birth, Berlin was a ruin, where the *Trummerfrauen*, the 'rubble-women', sifted the debris with their bare hands. The city's suffering continued, for post-war hostilities between the victors first rendered it an island in the prison-state of East Germany, then threw a concrete wall across its centre, with death as the punishment for crossing it. Again it was on Berlin's fateful date, 9 November, when the wall was breached in 1989. Now Berlin is once more the capital of a revitalised Germany, and the Reichstag is alive again, but its past is more haunted than that of any other capital city, and the question of what to restore and how is fraught with symbolic significance, awakening both pride and shame in its people. Berlin will perhaps survive and flourish only by becoming less different, less of a symbol of the past, and more like any other city in western Europe.

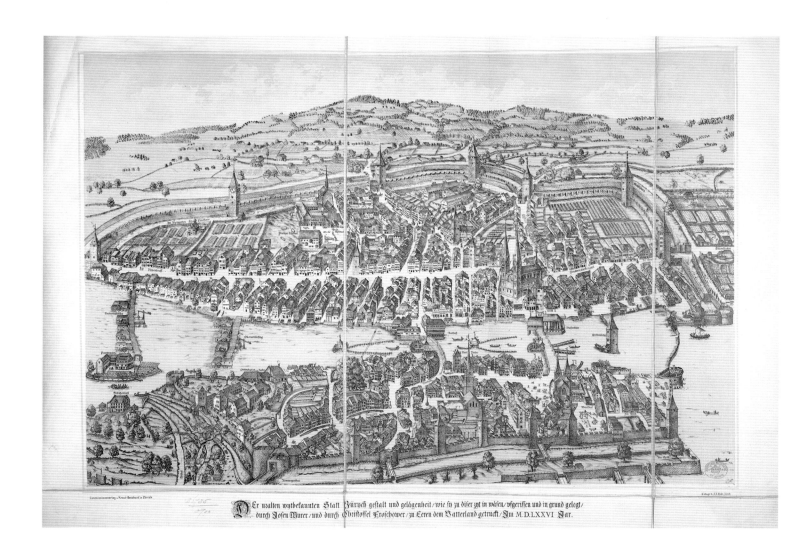

Der uralten wytbekannten Statt Zürych gestalt und gelägenheit / wie sy zu diser zyt in wäsen / vßgeriſſen und in grund gelegt / durch Joſen Murer / und durch Chriſtoffel Froſchower / zu Eeren dem Vatterland getruckt / Im M.D.LXXVI Jar.

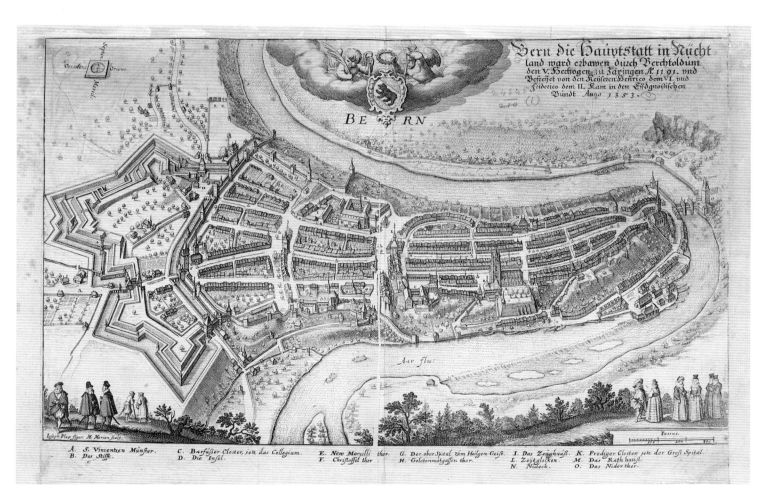

Bern die Hauptſtatt in Nücht land ward erbawen durch Berchtoldum den V. Hertzogen zu Zäringen A. 1191. und Befreyet von den Keyſeren Henrico dem VI. und Friderico dem II. Kam in den Eydgnoſſiſchen Bundt Anno 1353.

BERN

Occidens — Oriens.
Septentrio. Meridies.

Aar fluß.

Ioſeph Plepp figur: M. Merian ſculp.

Paſſus.

A. S. Vincentzen Münſter.   C. Barfüſſer Cloſter, ietz das Collegium.   E. New Marcilli thor.   G. Der ober Spital zum Heilgen Geiſt.   I. Das Zeughauß.   K. Prediger Cloſter ietz der Groß Spital.
B. Das Shifft.   D. Die Inſel.   F. Chriſtoffel thor.   H. Golotenmatgaſſen thor.   L. Zeytglocken.   M. Das Rathhauß.
N. Nideck.   O. Das Niderthor.

Bern and Zurich

Bern, the seat of the federal parliament, and Zurich, the financial and artistic centre, vie for the position of unofficial capital city of Switzerland. Bern is picturesque, a piece of living history, staid and civilised, intensely Swiss. Zurich is cosmopolitan, bohemian, a meeting place for international ideas and international finance. What both cities have in common is the beauty of their setting: Zurich at the river outfall from a spectacular mountain lake, and Bern almost an island, high up above the river which forms a magnificent loop around the city.

Zurich is much older, with evidence of prehistoric dwellings on the margins of the water. The narrow mouth into the River Limmat offered the only crossing point along the 20-mile (32-km) length of the lake, and the Limmat flowed on to join the Aare and then the Rhine. In Roman times *Turicum* was a bridge settlement rather than a town proper. Medieval Zurich was concentrated about the two bridges, the Munsterbrucke and the Rathausbrucke, and, unusually, development was fairly evenly spread across the two banks, with important buildings, churches and gates on both sides. Watermills proliferated in the midstream, as can be seen in this vivid engraving of 1576. From 1518 to 1535, Zurich was the source and centre of the Reformation in Switzerland, and Huldrych Zwingli was its leading figure, with his preaching in the Grossmunster; although he never founded a church, Calvinism eventually triumphed in the city. Tension between the Protestant city and the Catholic countryside continued to provoke armed conflicts, and as late as the mid-nineteenth century it was still possible that the Swiss Confederation might split apart over religion. The image of Switzerland as a land of peace, neutrality and aloofness from conflict is entirely a modern one. The policy of neutrality emerged not from some ideal of pacifism, but from the realisation that Switzerland, divided by language and religion, could not survive if it became entangled in European conflicts. While neighbouring states made war, neutrality made Zurich a safe home for money and for exiled intellectuals, such as Lenin and James Joyce, and the founders of surrealism.

It was in 1848 that Bern was chosen as the federal capital. While the architecture of Zurich has evolved steadily throughout the nineteenth and twentieth centuries, in the Middle Ages the people of Bern had found a form which was to endure essentially without change. This was the pattern of streets running east–west along the peninsula, radiating out from the single bridge, the Nydeggbrucke to the east, and intersected by open spaces. The point where the peninsula began to widen offered a natural site for a single short wall which could protect the entire city. Along the east–west streets were built Bern's famous *Lauben*, the arcaded walks fronting the innumerable shops. These were originally of wood, but after a devastating fire in 1405 which destroyed more than half of the city, they were rebuilt in stone on an identical pattern. Here the shopper can linger for hours immune from sun, wind, rain or snow - surely the invention of a retailing genius. The charm of the *Lauben* is increased because the arcaded walks are not rigid and geometrical, but curve gracefully, following the natural contours of the peninsula. Nor are they narrow alleys, but are broad enough to admit fountains and statues. Bern was for a long time a small community, confined to these streets only - in 1830 the population was still only 20,000. It was in a sense an 'Ideal City', but one which had evolved naturally and not on the drawing-board of a Renaissance architect. Bern has since expanded far beyond its original boundaries, but the character of the *Altstadt* is still largely intact.

Above: Zurich by Murer and Froschauer, 1576, looking from west to east across the River Limnat, with the lake out of sight to the right. The famous town hall built on the bridge is in the centre, and in the background is the cathedral church, the Grossmunster, from which Zwingli had directed the Swiss Reformation.
The British Library, Maps 26735 (13)

Below: Bern, 1653. Matthaeus Merian's classic view from the south shows perfectly the compact old city, tightly confined in the loop of the River Aare. The famous arcaded streets run east–west, and only the neck of the peninsula needed to be heavily fortified.
The British Library, Maps 24845 (1)

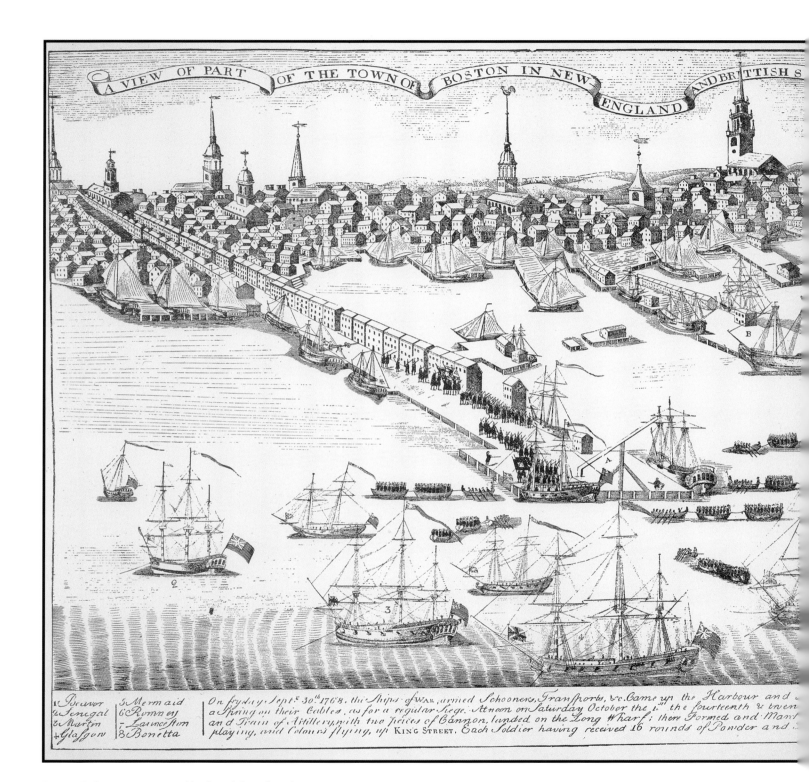

A VIEW OF PART OF THE TOWN OF BOSTON IN NEW-ENGLAND AND BRITTISH S

1 Beaver	5 Mermaid	On fryday, Sept.r 30.th 1768. the Ships of War, armed Schooners, Transports, &c. Came up the Harbour and
2 Senegal	6 Romney	a Spring on their Cables, as for a regular Siege. At noon on Saturday October the 1.st the fourteenth & twen
3 Martin	7 Launceston	and Train of Artillery, with two pieces of Cannon, landed on the Long Wharf; there Formed and Mar
4 Glasgow	8 Bonetta	playing, and Colours flying, up KING STREET. Each Soldier having received 16 rounds of Powder and

Boston Harbour, 1768, engraved by the celebrated Paul
Revere. The wharf continues the line of Great Street, and
extends a quarter of a mile into the harbour. Later, more
wharfs were added, which were joined by lateral landing
stages, and the in-filling of the spaces between them began
Boston's long process of reclaiming land from the sea.

The British Library, x802/5055

Boston

The historic heart of New England, Boston was for generations America's closest link to its European heritage, but at the same time it became the earliest centre of American culture and the fountainhead of revolution and independence. Once that independence was secured, Boston's identity would once again centre on its pride at being the most European of American cities. The city stands on a peninsula once linked to the mainland only by a narrow strip of land on the western side. Much of the modern city is built on land steadily reclaimed over two hundred years from the sea and from the Charles River.

The first inhabitant of Shawmut, as the peninsula was called, was William Blackstone, a solitary survivor from an abandoned colony to the south. He apparently dwelled in isolation from 1625 until 1630, when he invited a group of Puritans living north of the Charles River to leave their unhealthy colony and settle on the peninsula. These people were members of the Massachusetts Bay Company, established by royal charter from King Charles I in 1629, whose governor was John Winthrop. For various reasons connected with events in England, the colony was left in peace for fifty years, to develop into a self-governing theocracy. These were the people who bought William Blackstone's 45 acres of pasture for 30 pounds – then, around 150 dollars – the pasture which was to become Boston Common. Blackstone himself soon departed to the more relaxed atmosphere of Rhode Island, while Winthrop spelled out his stern vision of his community's future: 'We shall be as a city upon a hill: the eyes of all people will be upon us.' The site was renamed Boston for the Lincolnshire town in England from which the colonists had come. Physically the new settlement was laid out with none of the regular design which would characterise later planned colonial towns. Instead, the land features of the peninsula, with its hills, marshes and rocky banks, determined where the first streets should be.

The natural site for wharfs and docks faced the east, and the principal road therefore ran from the land bridge down to the harbour, roughly where the Fitzgerald Expressway now runs. A Dock Square and a Market Square were sited where this road turned towards the waterfront, the first for handling cargo, the second for a citizens' market, fronted by the town house, and the governor's house nearby. Later, by 1710, this street was in effect extended out into the harbour by the building of the Long Wharf. Other smaller wharfs projected into the harbour and these were eventually joined, and their spaces in-filled, thus reclaiming land from the sea; today's waterfront extends well beyond the original waterline, so that Long Wharf is now no longer than the others. The old town naturally grew towards the eastern seafront, but to the north was a large marsh which was dammed and became known as the Mill Pond, because a spillway was cut towards the harbour to power corn- and saw-mills.

The city prospered through fishing, shipbuilding and trade with Europe, and became the largest community in British North America, the population rising from 3,000 in 1660 to 17,000 in 1740. The Puritan laws which governed Boston were sternly enforced: the unorthodox were driven out, or in the last resort executed. Prominent among the theocratic leaders were the Mather family who tried fiercely but, in the end, unsuccessfully to defend the religious independence of the Massachusetts colony when the British crown began to reassert control in the 1680s.

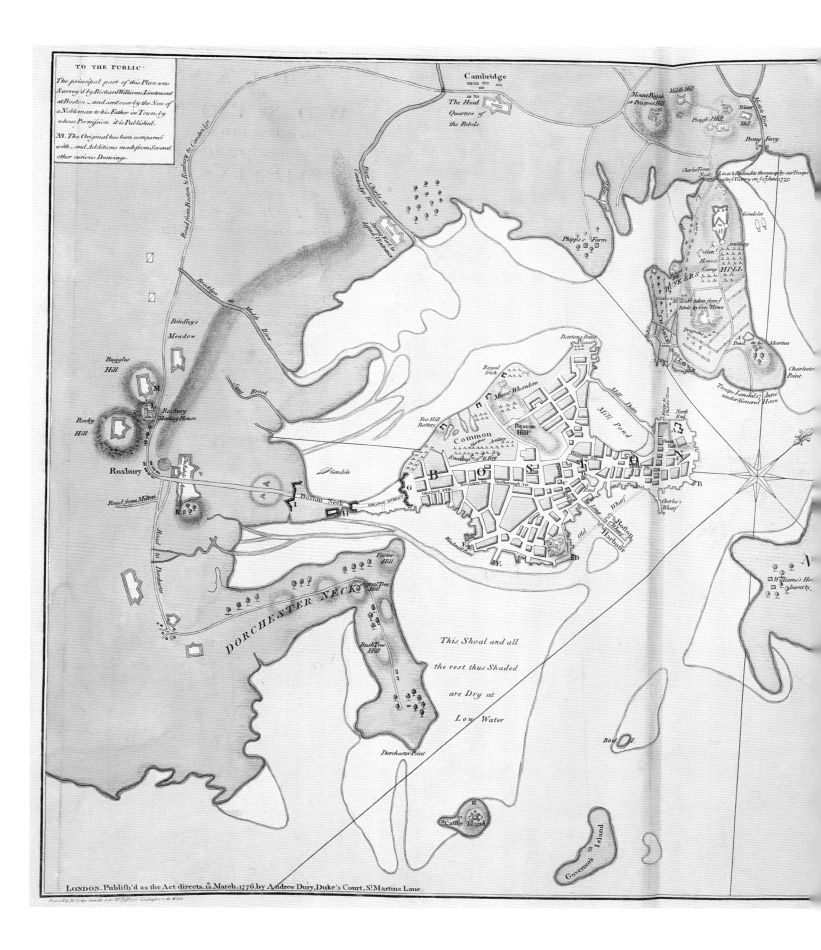

Cambridge

The Head
Quarters of
the Rebels

Mount Rigale
or Prospect Hill

Middle Hill

Plough'd Hill

Winter
Hill

Medford River

Penny Ferry

Charles Town
Neck

Lines & Redoubts thrown up by our Troops
after y Victory on y 17 June 1775.

Road from Boston & Roxbury to Cambridge

Charles River

Ferry Key to
Winter Encampance

Cambridge River

Gondolas

Gen.l
Howes
Camp

Artillery

BUNKERS

HILL

Redoubt taken from y
Rebels by Gen. Howe

Phipps's Farm

Dragoons

A Pond

Marines

Brookline

Mud River

Bindley's
Meadow

Charles Town

Ruggles
Hill

M

Troops Landed 17 June
under General Howe

Charlestown
Point

Royal
Irish

Bartons Point

Rocky
Hill

Roxbury
Meeting House

Royal Irish

Mount Wheredon

Mill Dam

North End

Long Brook

Fox Hill
Battery

Beacon
Hill

Mill Pond

A

Church Steeple

Roxbury

Gondola

Common

Artillery

Marines

Embodyes of 6 Key

B O S T O N

New Jersey Str.

Marlborough Str.

C

Pond

B

Road from Milton

I

Boston Neck

H

ORANGE STREET

ORANGE STR.

Charles's Wharf

Long Lane

Boston
Wharf

Boston Harbour

K

Road to Dorchester

Foster
Hill

Windmill Point

F

E

B

Signal Tree
Hill

D O R C H E S T E R   N E C K

Williams's Ho.
burnt by

N

Bush Tree
Hill

This Shoal and all

the rest thus Shaded

are Dry at

Low Water

Bird I.

Dorchester Point

Castle Island

Governors Inlet

LONDON, Publish'd as the Act directs, 12 March, 1776, by Andrew Dury, Duke's Court, St. Martins Lane.

A PLAN
OF
BOSTON,
and its ENVIRONS.
shewing the true SITUATION of
HIS MAJESTY'S ARMY.
AND ALSO THOSE OF THE
REBELS.
Drawn by an Engineer at Boston, Oct.r 1775.

REFERENCE.

A { Corpse Hill, a Battery of 8 Pieces of Cannon, Mortars, &c. &c. Erected to
favour the Troops Landing on y 17 June &, to set Fire to Charles Town ......

B { North & South Batteries, built by the Province for the defence of y Harbour;
they are in a ruinous State.

C. Town Hall ....

D. Fanuil Hall ....

E. Two Batteries Erected on Wharfs against Dorchester Neck .....

F. Fort Hill, a proper Place for constructing a Citadel ....

G. Fortification now constructing for the immediate defence of the Town .....

H. A Block-House & 2 Strong Batteries pointing on part of Dorchester Neck.

I. Lines to defend the Boston Neck .....

K { A Hill from whence the Enemy annoy y Centries & Officers with small
Arms, but seldom do any Execution .....

L { Roxbury Meeting House, upon a Hill from whence y Enemy often Fire Cannon
into the Lines .....

M { A Strong Post of the Enemy Fortified in appearance with great Judgment,
& much Elevated, from whence with a 24 Pounder they can just reach
the Lines .....

EXPLANATION.

The Works shaded Green, shew those constructed
by His Majestys Troops.

The Works shaded Yellow, shew those thrown up
by the Rebels, as they appear from Boston.

Road to Marble Head & Salem

Winnisimond Ferry

ES ISLAND

HOG ISLAND

Yards or Half a Mile

The Puritan isolation could not survive the economic openness which created Boston's wealth, but the spirit of independence was transferred to the political sphere, with the result that Boston became the nerve centre of resistance to overbearing government from London, and the site of the first flashpoint conflicts of the Revolution. The Boston Massacre of 1770, the Tea Party, Paul Revere's Ride, the Battle of Bunker Hill – all these have entered into the epic of North America's birth.

It was after these great events, whose locations may still be traced between the modern skyscrapers, that the classic Boston façade took shape. From the 1780s onwards, the slopes of Beacon Hill and Mount Vernon were built over with elegant town-houses, and the architect Charles Bulfinch brought back from his tour of Europe inspired ideas which produced the elegant Louisborg Square and Franklin Place. The levelling and improvement of the land now permitted a far more regular grid-pattern of streets to be laid out. By the 1850s, work at last began on the reclamation of the Back Bay area to the north-west, creating a far wider link with the mainland and encouraging the design of a wide, Parisian-style boulevard fronted by the imposing houses known as the Back Bay palaces. The built-up area of the peninsula had now probably quadrupled since 1700, and the vast changes brought by the wave of immigration from Europe were still to come. By this time Boston had been overtaken in economic terms by New York and Philadelphia, but in cultural terms it was still the home of the aristocracy. Emerson, Hawthorne and Thoreau created a school of intuitive individualism which would profoundly affect American literature. The atmosphere of elitism, intellectual and social, which pervaded nineteenth-century Boston is beautifully evoked in *The Education of Henry Adams*, and it hovers in the background of much of Henry James's fiction. This legacy is as much a part of what people seek in Old Boston as the eighteenth-century buildings and the monuments of the Revolution.

Boston in March 1776, showing the island status of the historic town. This plan was published in London by Andrew Dury to show the disposition of the British troops and the 'rebels', but it was out of date on the day it was printed. The 'rebels' had been encamped to the west and north during the year-long siege, and despite the initial British success on Bunker Hill above Charlestown, they were compelled to withdraw from the city by sea.
The British Library, Maps 1. Tab. 44 (17)

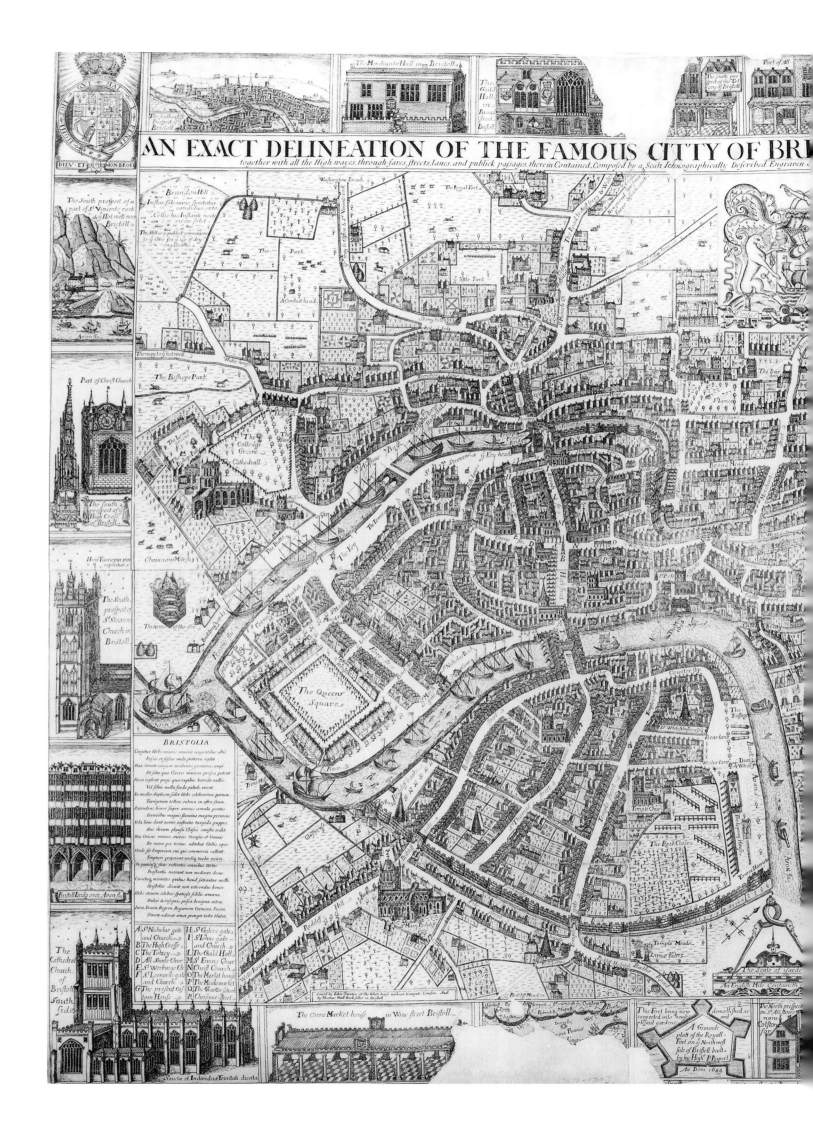

AN EXACT DELINEATION OF THE FAMOUS CITTY OF BRI
together with all the High wayes, through fares, streets, lanes, and publick passages, therein Contained, Composed by a Scale Ichnographically Described Engraven

The Merchants Hall in Bristoll.

The South prospect of a part of St Vincents rocks, & Hot well neer Bristoll.

Avon flu:

Part of Christ Church

The South prospect of y High Cross in Bristoll.

How Turris pat picta reperitur.

The South prospect of St Steeven Church in Bristoll

The Armes of the See

BRISTOLIA

Conspicua Urbe maris mures angusti bus altis
Fossæ et fossæ unda portarii replet
Has iterum congens incidentes geminos campi
Et fasta quæ Cireris munera prisca patent
Rura replent pagi quæ rupibus horrida vallis
Vel sileni nulla foeda palude recit
In medio duplicem sedet Urbs celeberrima portum
Tamquetum tollens culmen in astra suum.
Extendunt hunc super amnes æmula pontes
Fornicibus magni flumina magna prementes
Vela hinc dant ventis restrata turgide puppes
Hoc iterum plausu Classis onusta redit
Has Orents merces mercas Occasus et Omnis
Fit mare per terras advehit Orbis opes
Unde sit Emporium cui qui commercia callent
Emptores properent undiq turba viruin.
In pandig flux redeuntes omnibus Urbe
Prestantes narrent non mediocre decus
Cunctaq mirantes quibus haud satiantur ocilli
Bristolia dicunt non retinenda domos
Ulebs cernim celebres spatiosa fidelis amæna.
Dulci & insignis prisca benigna nitens
Jura Deum Regem Regionem Crimina Paces.
Serevet adorat amat protegit Odit Habet.

A	St Nicholas gate and Church
B	The High Crosse
C	The Tolzey
D	All Saints Chur
E	St Werburge the and Church
F	St Leonards gate and Church
G	The present Cus tom House
H	St Giles gate
I	St Johns gate and Church
L	The Guild Hall
M	St Ewins Chur
N	Christ Church
O	The Market house
P	The Meale market
Q	The Gramer school
R	Christmas street

The Cathedral Church of Bristoll South Side

Bristoll Bridg over Avon flu.

Brandon Hill

In flux sole novæ spectator candidus orto
Collis hæc Instans nota vacare solet.
The Hill is a public conveniencie to y Cittie for y use of dry ing Cloath.

The Park

A Conduit head.

The Royal Fort

Little Park

Stony hill

Hallie house

The Bishops Park

Frog lane

Hospitall

The Lemes

The Colledge Greene

The Cathedrall

The Butts

Chanons Marsh

Avon flu:

The Queens Square

Temple Chur

The Rack Closes

The West Almshouse

Beare lane

Water lane

Dunn Whart

Bridwell

Glashouse

Treen Mill

Raine

Reddiff Hill

St Mary Redliff

Redliff Head

Temple Meade

Lime Kills

Temple Meads

THE SCALE OF YARDS

An English Mile Contain th

Loud by John Ouerton at the White horse weithout Newgate London And by Thomas Wall Book seller in Bristoll

The Corne Market house in Wine street Bristoll.

Sancta et Individue Trinitati dicata

A Grounde platt of the Royall Fort on y North west side of Bristoll built by the High P. Rupert. An Dom 1644

The North prospect of All nons Colston

DIEV ET MON DROIT

Bristol

The fortunes of this medieval wool town differed from so many other similar towns because its river, the Avon, although not large, provided a navigable link with the sea, the Severn Estuary lying just seven miles downstream. By 1200 Bristol was exporting woollen cloth to Europe and importing wines from Bordeaux and Spain. So great was its trade that new quaysides became necessary, and massive works were carried out around the year 1250 to divert the Avon and the smaller river, the Frome, into new channels, thus creating a virtual island in the town's centre. A medieval chronicler wrote: 'Bristol is a city nearly the richest of all the cities in the country, receiving merchandise by sailing vessels from all foreign countries ... a great and strong tide ebbing and flowing abundantly night and day, causes the rivers on both sides of the city to run back upon themselves into a wide and deep sea; forming a port very fit and safe for a thousand vessels, it binds the circuit of the city so nearly and so closely that the whole city seems to swim on the water and sit on its banks.'

This island effect is now lost because the Frome has been channelled entirely beneath ground. Bristol was for several centuries England's second city, and it was the peculiar boast of its merchants that they alone kept their trade independent of London, importing goods from America and Europe and reselling them throughout the West Country, where the rest of England's foreign trade passed through London. The Bristol merchant fleet was to play a considerable role in the history of exploration: the mysterious Venetian-born John Cabot sailed from here in 1497, financed by a local syndicate, and discovered Newfoundland, bringing back reports of the teeming fishing-grounds around the island. On his second voyage a year later, Cabot and his fleet vanished from the pages of history, but within a few years Bristol seamen were regularly crossing the Atlantic to take cod from the Grand Banks. A 'Society of Merchants Venturers' was formed in the city, but no further historic discoveries resulted. In the seventeenth century Bristol's fortunes became deeply involved with the 'triangular' Atlantic trade: Bristol ships took slaves from West Africa to the Caribbean, and later to the American colonies, returning to England with sugar, cocoa, tobacco and cotton. The wealth from this period created Bristol's elegant squares and crescents and the fashionable suburb of Clifton. By the early nineteenth century, though, the rise of the Lancashire cotton industry and the abolition of the slave trade removed much of this commerce to Liverpool, where the Mersey was also better adapted than the Avon for large ships. In response, new docks were constructed nearer the sea at Avonmouth and Portishead, and the tidal waters of the Avon were diverted to leave a tideless harbour in the city with a constant depth of water. Despite these strategies, however, Bristol's maritime trade shifted permanently to the Severn Estuary, and ebbed away from the city itself.

Although famous as a seafaring town, Bristol's literary history is also distinguished. It was in a Bristol tavern that Daniel Defoe

Bristol: James Millerd's map of c.1710. A lively depiction of what was then England's second city, compact, peninsula-like, crowded with shipping. Among the border pictures of the town's churches is an interesting early view of the Avon Gorge. The map is a reminder of the time when the junction of Corn Street and Broad Street was the heart of Bristol.
The British Library, Maps K Top 37, no.32

met Alexander Selkirk, after which he based the character of Robinson Crusoe on Selkirk's account of his years marooned on the island of Juan Fernandez. Thomas Chatterton knew the church of St Mary Redcliffe well, and it was in the church chest that he claimed to have discovered his forged medieval Rowley poems. Most important of all, it was the Bristol bookseller Joseph Cottle who befriended the young Wordsworth and Coleridge and who published their *Lyrical Ballads* in 1798, thus initiating a revolution in English poetry. Curiously, Bristol holds a special place in the history of Portuguese literature, for it was here that the great novelist Eca de Quieros lived when he was the Portuguese consul during the 1880s, and here that he wrote some of his best works, including *The Sin of Father Amaro*, a classic of nineteenth-century realism.

Bristol: a view of Broad Quay around 1780, showing the way the docks penetrated into the very heart of what was then England's second city.
Bristol City Museum and Art Gallery/www.bridgeman.co.uk

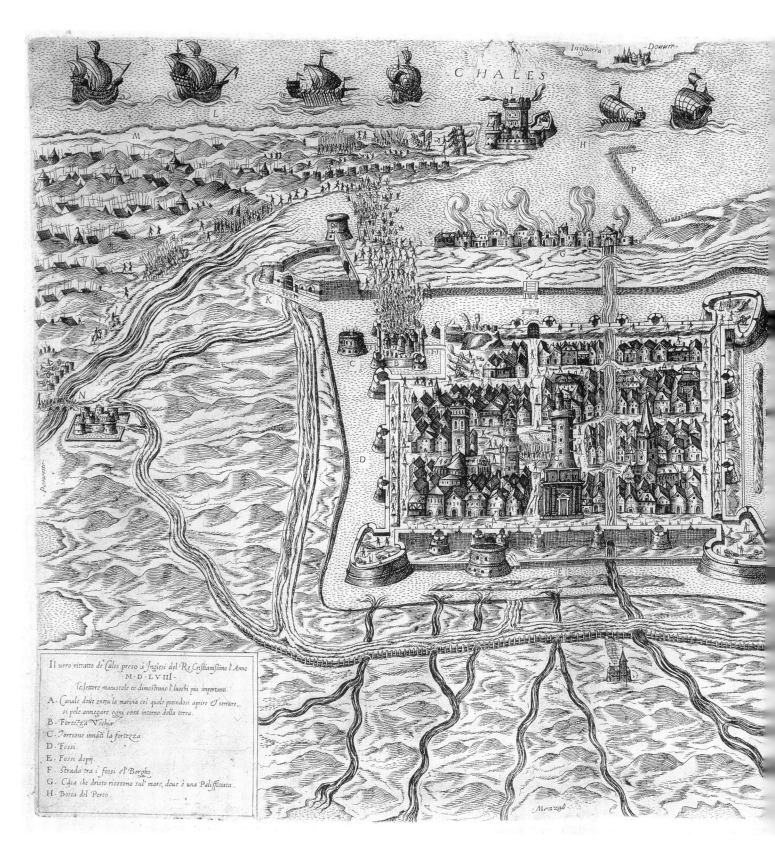

Calais, 1558, by Florimi. A dramatic view of the battle in
1558 in which France seized back England's last foothold
on the French coast, which had been occupied for over
two centuries. The harbour is well forward to the sea,
while the town is effectively a fortified island.

Calais

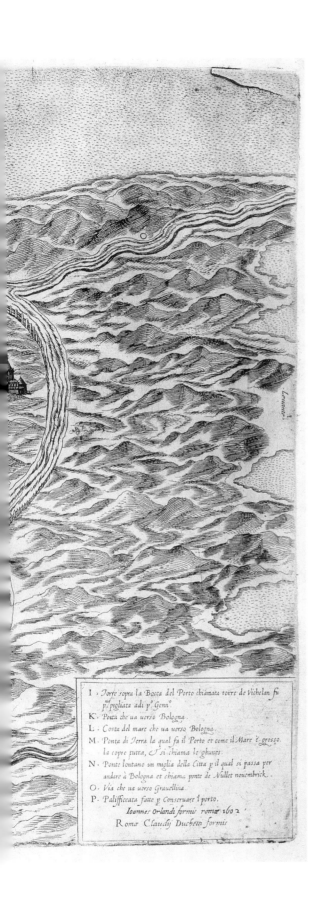

If there is a town in France which is a part of England and of English history, it is Calais. Some ten million English visitors pass through it each year, and the imposing tower of the Hotel de Ville, graceful in spite of its top-heavy design, is a landmark pointed to by thousands from the approaching ships. The old town is now entirely an island amid its many harbour basins. Originally a fishing village, it was so substantially walled in the mid-thirteenth century that in 1346–47 it withstood a ten-month siege by the English army of King Edward III. The siege was ended by the surrender of the six burghers, who offered their lives to save the town – the episode celebrated in Rodin's famous statue. In the event, Edward spared the six men after the intercession of his French wife Isabella, but then drove out the French inhabitants and turned Calais into an English base, from which the Hundred Years' War would be conducted.

Calais was called 'the fairest jewel in the English crown' until it was re-taken in 1558, at the siege led by the Duc de Guise, pictured here. Its loss ended two centuries of English territorial pretensions in France, and it proved a fatal blow to Queen Mary Tudor, who famously died with the word 'Calais' etched in her heart. Thereafter, Calais and its hinterland were long known as the 'Pays Reconquis'. The historic roles of England and France were reversed in the Napoleonic era, when part of the French army camped here for many months in 1805, preparing the invasion of England which never materialised. Calais's strategic link with England was again in evidence in 1940, when it became the main objective in the German drive for the sea, and later when it was used as the launching base for V-weapon attacks on England.

More Flemish than French in character, Calais today has the hectic air of a place of transit. It is once again besieged by the English in their thousands, while the people of Calais are cynically resigned to their historic fate as an outpost of England.

Calais: the most famous incident from the town's history: the surrender of the six burghers to King Edward III, from a medieval manuscript.
Lambeth Palace Library

Cambridge

According to legend, Cambridge University, which forms the heart of the city of Cambridge, was born in the early thirteenth century when a group of dissident scholars decided to abandon Oxford and found a new seat of learning of their own. The great difficulty with this story is that it fails to explain why they migrated almost 100 miles (160 km) to the little market town on the edge of the fens, unless there were already scholars settled there, that is, unless an embryo university already existed. The priority dispute between Oxford and Cambridge remains a harmless sport. By the end of the thirteenth century, groups of scholars had already begun to congregate into the halls which would form the nuclei of the later colleges. St Peter's Hall (Peterhouse) was the first, and others followed, steadily acquiring land for their schools, chapels and hostels along the east bank of the River Cam, and the university soon came to dominate the medieval town. The major event in this process was Henry VI's decision to found his magnificent new King's College, for which he bought land, demolished old houses, and effectively refashioned the city centre. Further colleges were formed at the Reformation out of the suppressed religious houses - Emmanuel, Sidney Sussex and Magdalene - while Henry VIII re-founded and enlarged Trinity. As in Oxford, many of the colleges have been rebuilt or extended in the classical style.

Despite the formal similarities between the two universities, Cambridge has acquired a rather different identity from that of Oxford. For reasons good or bad, Oxford is seen as a place of poetry, of contemplation, of aesthetic ideals, of elitism and idleness. Cambridge is seen as a place of science, of criticism, of commitment, of intellectual austerity. Leaving aside the awkward fact that Cambridge is decidedly richer in major poets than Oxford - with Spenser, Marlow, Milton, Dryden, Gray, Coleridge, Wordsworth, Byron and Tennyson - this view seems to have a certain basis in fact. Science has long occupied a central place in Cambridge, dating from Newton's period as student and teacher. Much later, the work J.J. Thomson and Ernest Rutherford made the Cavendish laboratory the historic centre of modern atomic physics. It was at Cambridge, too, that Crick and Watson unlocked the secrets of DNA and inaugurated a new era in biological science. In the years 1910 to 1925, it was Bertrand Russell and then Wittgenstein who redefined the task of philosophy as the logical analysis of language, while Keynes placed economics on a new scientific level. In literature, I. A. Richards and F. R. Leavis sought to establish a new and rigorous approach to literary analysis, scorning the dilettantism and gossip of Bloomsbury and the rest of literary London. Leavis, with his moral ideal of literature, always insisted that the English faculty was the heart of Cambridge - a fantastic claim when one considers the achievements of Cambridge science. It was this claim which led to the celebrated 'Two Cultures' controversy with C.P. Snow in 1959, in which two aspects of Cambridge criticism clashed head-on.

Yet Bloomsbury was largely formed at Cambridge, and E.M. Forster explained the blend of emotional and intellectual excitement which was the keynote of the group: 'As Cambridge filled up with friends it acquired a magic quality ... People and books reinforced one another, intelligence joined hands with affection, speculation became passion, and discussion was made profound by love.' The aestheticism and homosexual friendship praised here formed the background to the spy scandals of the 1950s. Burgess, Maclean, Philby and Blunt all met there, all imbibed Forster's ideals, and all embraced Marxist politics, with the results that we know. Was this a fulfilment or a perversion of Cambridge's critical philosophy? Do the Oxford and Cambridge stereotypes really stand up to examination?

Cambridge in 1574 by Richard Lyne. A delightful plan of old Cambridge, its layout essentially unchanged today, but lacking the classical buildings which we now know. This engraving illustrated a polemical book which set out to prove that, contrary to received wisdom, Cambridge University was older than Oxford's.
The British Library, Maps C24.a.27

Left: Cape Town and its anchorage in 1675 seen from the sea, with the distinctive Table Mountain and Devil's Peak rising behind. The small Dutch colony was one of the most celebrated maritime gateways in the world.

The British Library, Maps C.8.b.13 (25)

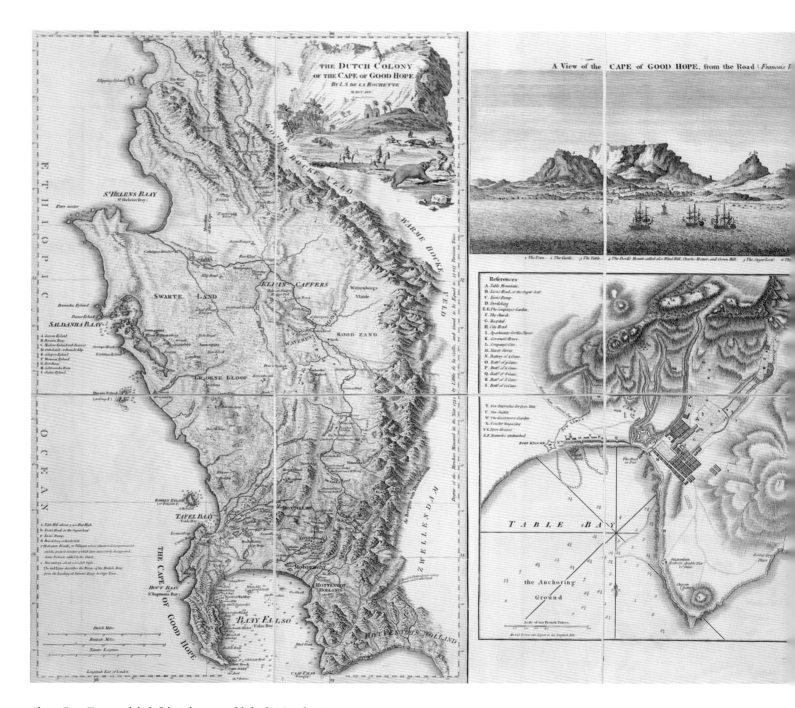

Above: Cape Town, a delightful triple map published in London in 1795 by William Faden, when the region was still a Dutch colony. The general map shows well the rugged, mountainous hinterland of the Cape; to the right is the unmistakable view from the sea of the spectacular Table Mountain, and the still tiny township with its citadel.

The British Library, Maps C.23.e.5 (3)

Cape Town

Cape Town is the oldest European settlement in southern Africa, and the cape from which it takes its name became one of the great landmarks of the maritime world, the gateway to India and the Far East. Strangely, the cape was not at first seen by Bartolomeu Dias, the first European to round it in 1488, because storms had driven him far out to sea; only on his homeward voyage did he sight the rocky peninsula and the inviting bay which lay to the north, over which loomed the distinctive flat-topped mountain. He christened it 'Cape of Storms', but the Portuguese king, realising its significance, chose the more up-beat 'Cape of Good Hope'. Over the following 150 years, Portuguese, English and Dutch mariners anchored in Table Bay, taking on fresh water and bargaining with the natives for supplies of meat and fruit. Francis Drake called it 'the fairest cape in the whole circumference of the earth', and in 1620 an English party raised a flag and claimed the territory for their king, an empty gesture without sequel.

Not until 1652 was a permanent settlement established by the Dutch East India Company. Jan van Riebeeck with around 100 men was commissioned to build a supply base for the Company's ships en route to and from the Indies, which they did in the face of fierce native hostility. Within two years, fruit and vegetables were being grown, while vineyards were planted on the mountain slopes. Just four years later, the Company began to release men from its employ to become free farmers and burghers, setting Cape Town on course to becoming a self-governing colony. In the 1670s a substantial new bastion on the European model was completed, and around it a simple rectangular grid of streets. The main thoroughfare down to the seashore was named the Heerengracht, after the canal in Amsterdam; it was renamed Adderley Street in the mid nineteenth century. Water courses were dug down from the mountain to replenish the ships. In the late 1680s the colony was enriched by the arrival of several hundred French Protestant refugees, who were granted land, and who developed the vineyards and the wine trade. The native population, christened Hottentots by the Dutch, were severely affected by outbreaks of smallpox, and the Dutch imported slaves from other parts of Africa and from the East Indies. By 1750 the town reached back from the shoreline to the slopes of Table Mountain, with many fine houses in the Dutch gabled style, populated by around 15,000 Europeans, with a similar number of slaves.

The turning point in Cape Town's history came in January 1806, when a British force took the city in order to thwart possible French designs on sea routes to the east, and on India itself. The city soon became an important building-block in the edifice of the British Empire, a naval supply base for the Indian trade. The Anglicisation of Cape Town in the early nineteenth century had far-reaching effects. Slavery was abolished in 1834 and rights were granted to people of mixed race, these events provoking the Dutch exodus of the late 1830s to found the Boer Republics in the north. On a practical level, the modern port was founded with the building of the western breakwater to protect the anchorage from the Atlantic gales. Rail lines to the interior of the Cape Colony were built in the late 1850s.

The importance of Cape Town might have declined dramatically with the opening of the Suez canal in 1869, but that event coincided exactly with another – the discovery of gold and diamonds in the Transvaal, which revitalised the economy of Cape Town. The dominant figure of the late nineteenth century was Cecil Rhodes, with his grandiose dreams of imperial Africa, a dream which created the conditions for the Boer War, during which Cape Town became the centre of the British military operation. The visible legacy of Rhodes's presence is the Prime Ministerial residence of Groote Schuur, where van Riebeeck's 'great barn' had once stood. In 1910, when the Union of South Africa was formed out of the merger of the Cape Colony with the Boer Republics, Cape Town became the parliamentary capital.

Throughout the grim apartheid years, Cape Town maintained a reputation for resistance to the government's drive towards racial segregation, a resistance focused in the Anglican cathedral of St George, standing immediately opposite the Parliament building. Yet just a few miles out in Table Bay lay Robben Island, once a leper colony, then later the site of the notorious prison where political opponents of the regime were confined. With the revolutionary changes of 1994, Cape Town assumed a new historic role as home to South Africa's first multi-racial government.

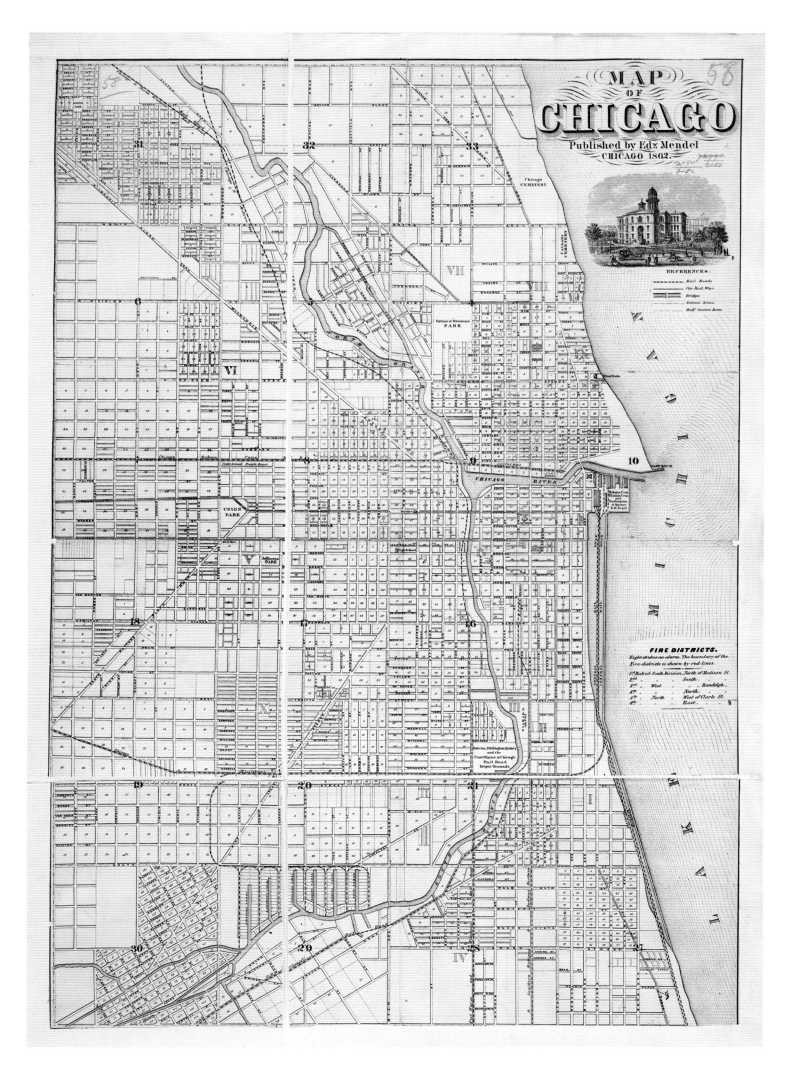

MAP
OF
CHICAGO
Published by Edw. Mendel
CHICAGO 1862.

REFERENCES:

FIRE DISTRICTS.

Chicago

'The city of speed', the 'I will' city, and 'the city where you don't have to wait for the future' – these were some of the characteristic boasts of the people of Chicago in the hectic decades of the late nineteenth century, and no greater contrast could be imagined between the religious or social idealism which lay behind the first New

BIRD'S EYE VIEW OF THE WORLD'S COLUMBIAN EXPOSITION

Above: Chicago's hosting of the 1893 World's Columbian Exposition symbolised the city's importance in the American economy.
The British Library, 0848.pp.49

Left: Chicago, *c.*1857. Conceived and planned by speculators as a settlement where the Mississippi trade routes would connect with the Great Lakes, the gridiron division of land plots was the classic way in which such a town could be built and sold.
The British Library, Maps 72787 (4)

England cities and the economic explosion which created Chicago. It was born suddenly, out of the plan to connect the Great Lakes with the Mississippi via a great canal. Such a canal had been dreamed of as early as the seventeenth century by the first French explorers, who saw it as forming the vital artery to unify New France. Much later, in 1822, the United States Congress authorised the state of Illinois to acquire a great swathe of land where a canal could be dug to link the Chicago and Illinois Rivers. After a further ten years of planning, construction began in 1832, and it was then that the first real estate plots were sold to settlers and speculators. It was the promise of the canal which drove the whole project forward, for the site itself was then an uninhabited, insect-ridden mudflat where a small river

emptied into Lake Michigan. The rigid gridiron street pattern was the classic form in which such speculative land sales were organised, and a mania for buying began which attracted 4,000 settlers in the first two years. Properties and farms outside the town were also traded, and the boom affected the entire region. It was said that property values could increase tenfold in a single day, between the 'morning buyer' and the 'evening buyer'. As one commentator remarked, 'Where a stream, however diminutive, found its way to the shore of the lake, the miserable waste of sand and fens, which lay unconscious of its glory, was suddenly elevated into a mighty city, with a projected harbour and lighthouse, railroads and canals ... Not the puniest brook on the shore of Lake Michigan was suffered to remain without a city at its mouth.'

Inevitably this boom periodically collapsed and then revived, but when the canal was completed in 1848, the population had reached 20,000. Just as in Europe, however, the canal became virtually obsolete almost at once, for the first railways arrived in Chicago in the same year and within a decade the city was the focal point of a powerful rail network, linking Chicago to the cities of the east and to the sources of food, timber, coal and minerals which lay to the west. The population rocketed to 100,000 in 1860 and to 500,000 by 1880. Such intense change could not be achieved without trauma. Overcrowding and squalor were endemic. The great fire of 1871 wiped out most of the central area, only the landmark water tower surviving the inferno (this tower was part of a remarkable engineering feat, for it was a pressure-relieving valve for the huge pipe which pumped fresh water into the city from two miles out in Lake Michigan). Labour disputes were frequent and violent, culminating in the anarchist bombing in Haymarket Square in May 1886, and the violent strike at the great Pullman works. But still the city grew, always extending its rigidly square-patterned streets to the north and west. In the 1890s the world's first skyscrapers were conceived, to relieve pressure on space: the skeletal steel frames of the new buildings made heavy masonry walls unnecessary, and upward growth became possible. This architectural revolution spread to other cities and swiftly became the definitive symbol of urban America. Immigrants from Poland, Sweden, Ireland, Russia and Italy were drawn in by the

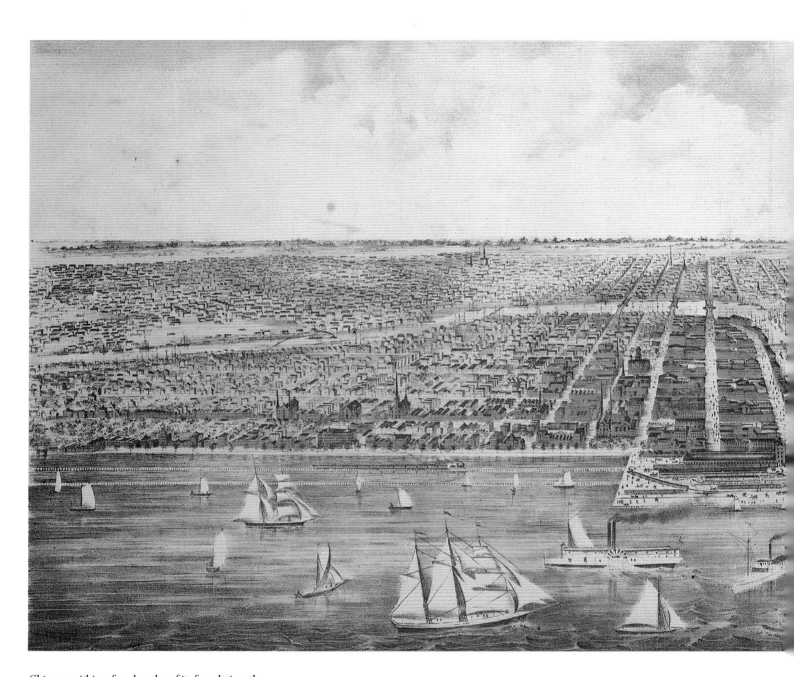

Chicago: within a few decades of its foundation, the
advent of the railways had transformed Chicago into
the massive lakeside port seen in this 1892 view.
The British Library, Maps 72787 (1)

thousand, to process the grain, the meat, the coal, the steel and the timber. By 1900 the population numbered 1.6 million, seventy-seven per cent of them of foreign birth. This was the setting of Upton Sinclair's sensational 1906 novel *The Jungle*, which exposed the brutality of the dynamic, unregulated industries on which Chicago's prosperity was built.

The city's arrival as the giant of the American economy was symbolised by the staging there of the 1893 World's Columbian Exposition, for which it outbid Washington and New York. This fair was planned on a gigantic scale: on a 550-acre site south of the city, the construction of a vast 'white city' of exhibition halls employed 40,000 people for almost two years. The buildings were in the neoclassical style, and some critics regretted the lost opportunity to showcase the new architecture. In six months 26 million visitors passed through the fair, almost half of the population of the United States. The economic impact on Chicago was massive, and one of the great pavilions was retained on the original Jackson Park site and now houses the Museum of Science and Industry. In the aftermath of the fair, a vision of the future city was drawn up in the famous Burnham and Bennett plan of 1909, which proposed to transform the functional city into 'the city beautiful'. Lakeside parks and drives, boulevards, a rim of forested greenbelt, new bridges, open squares – many of these features of the plan became reality between 1909 and 1939. This transformation represented the positive side of Chicago's development during a period when the city became notorious for gangsterism and corruption. The volcanic history of Chicago between 1850 and 1950 would be conceivable only in America. The ceaseless tide of immigrants, the rail network, the supplies of food and raw materials from the hinterland – all these came together to produce a level of growth that was phenomenal even by American standards, but which in the end crystallised into something lasting, massive and in its way elegant.

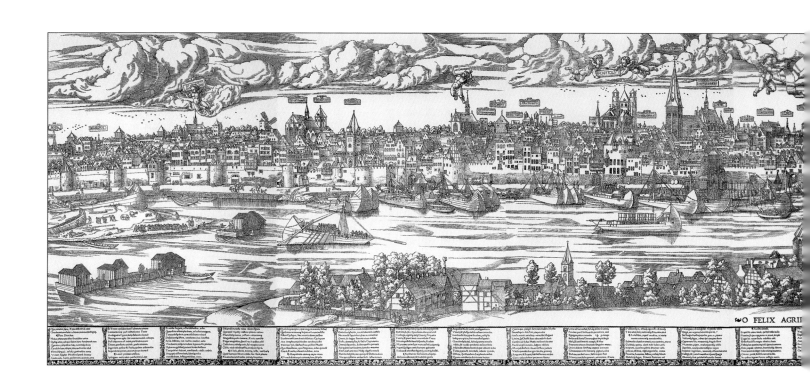

O FELIX AGRI

Cologne

Cologne today is overwhelmingly modern, but no city in Germany has more ancient roots. A Roman town, whose name means simply 'colony', it occupies a pivotal crossing point between Rhine traffic from the heart of Europe, and the east-west route from France to Germany.

mercial city in Germany, for the Rhine was navigable by small sea-going vessels. It was also a formidable fortress, with walls 3 miles (5 km) around on the landward side, punctuated by ten gates and guarded by numerous watchtowers; the river-frontage was almost a mile in length.

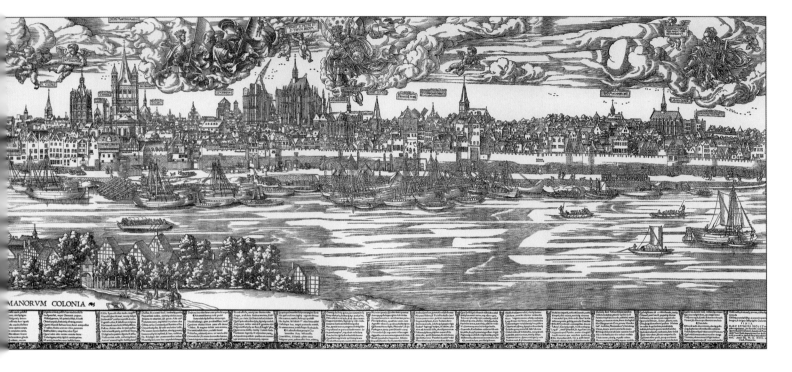

MANORVM COLONIA

Above: Cologne, Woensam's great panorama of 1500 does not reveal the clear, semi-circular layout of the city, but it conveys beautifully the vaunting architecture and the intense commercial activity of the riverside city, the crossroads of so many medieval trade routes. The great cathedral is seen under construction.
The British Library, Maps 149.e.28 (2)

Left: Nineteenth-century Cologne with its semi-circular form on the river banks of the Rhine. German school, artist unknown.
Archives Charmet/www.bridgeman.co.uk

The form of the city was simple and remained essentially unchanged: it was a semi-circle whose base was the riverside. The main streets were in a rectangular grid, bounded by the encircling walls. But as the walls were replaced and the city limits extended, concentric rings were formed. Such extensions occurred three times from the twelfth to the seventeenth century. Even the Rhine river-front was fortified, but nevertheless the city was almost destroyed by Norsemen in the year 881. The Emperor Constantine had constructed a substantial bridge on stone piles in 310 AD, which may have been destroyed in this attack – it did not exist in the Middle Ages. From time to time bridges of linked boats spanned the river, across to the small settlement of Deutz, but they were always vulnerable to the Rhine's powerful current. Not until the 1870s was a permanent bridge established – the Hohenzollern railway bridge.

By the thirteenth century Cologne, a free imperial city, had become the most thriving com-

The panorama by Woensam of 1500 shows the late-medieval magnificence of Cologne, its quays thronged with craft, and its towering churches watched over by saints and angels. The university was founded in 1300, and was home to the greatest theologians of Europe, Albertus Magnus and St Thomas Aquinas. The great landmark of Cologne has long been its vast cathedral (although the twin towers were not finally completed until the late nineteenth century), traditionally the resting place of the bones of the three magi, gifted to the city by the Emperor Barbarossa.

Cologne's prosperity declined somewhat in Europe's maritime age, but its central location made it once again a focus of the industrial revolution, and every rail journey in northern Europe seems to pass through Cologne, where the east-west routes cross the north-south. The city was devastated in World War II, and the cathedral now stands as a splendid survivor among the overwhelmingly modern architecture.

PERA

Sautari

Turquia

Porta delmeso

Sāt demeš

Sāt geor̄ gius

CONSTAN=
TINOPOLIS.

chiramos

palaciū Imparoris

Sēs Iohēs ō petra

portus olim
palaciī im
peratoris

porta

ypodromio

Calchedona

Sēs Iohēs ō
Insula

Planga

Portus sed destruct
precepto tenutorū

Constantinople

Constantinople (originally Byzantium, now Istanbul) is one of the great historic cities of the world, with a past no less eventful than that of Rome itself. A Greek settlement from the seventh century BC, it was successively occupied by Persians, Macedonians and Romans, before the Emperor Constantine proclaimed it his capital city and the 'New Rome' in 330 AD. Repeatedly ravaged by fire, earthquake, plague and human enemies, it nevertheless clung to its role as the centre of eastern Christendom, until the year 1453, when it finally fell to the conquering Turks under Sultan Mehmed II. It was a strategic bridge between Europe and Asia, a gateway city through which all trade from the Orient passed. It was the symbol of the Greek Christianity which became increasingly estranged from that of Rome, and it was constantly threatened by the neighbouring forces of Islam, although its greatest disaster was to be sacked and pillaged by the crusader army in 1204 and to languish under Latin rule for a further sixty years. After it had fallen to the Turks, the city became the capital of the Ottoman Empire. Its zenith was probably under the Emperor Justinian in the sixth century, when a series of magnificent buildings was created, and the population probably reached half a million.

The earlier of these two maps is a delightful manuscript illumination of the fifteenth century by Cristoforo Buondelmonte. The ancient walls, the church of Hagia Sophia and some imperial ruins such as the Hippodrome are instantly recognisable, but in reality the city at this date was in a state of decay, never having fully recovered from the years of hostile Latin rule. The district of Pera across the Golden Horn, with its landmark Galata Tower, was largely inhabited by Genoese merchants and was far more prosperous at this time. This view of the city first appeared in a guide to the ports of the Mediterranean composed in 1420, when the Turks were actually preparing to attack the city. That siege of 1422 failed, but the city's eventual fate was inevitable, and by the time this later copy of the manuscript was made in 1482, Constantinople had been part of the Turkish Empire for many years and the map was seriously outdated.

An image of the city in its new role as the Turkish capital is found in the great atlas of town plans *Civitates Orbis Terrarum* published by Braun and Hogenberg in 1572. The outstanding new landmarks are now the Topkapi Palace in the foreground, built in the 1460s, and the Suleyman Mosque of the 1550s, just to the rear of the centre. The portrait gallery of Ottoman rulers appears almost like a set of royal seals of the new ruling dynasty. The Arsenal or shipyard is misplaced, for it really lay on the Golden Horn, and this mistake was copied in many views of the city. We can only guess how the data for this evocative map was obtained. The Ottoman Turks had developed their own rich tradition of topographic art and, despite deep cultural hostilities, some slight trading contacts between Turks and Europeans were sustained; somehow Braun and Hogenberg must have been able to obtain the necessary sketches of the rebuilt city upon which to base their map. It was published at the very height of Ottoman power, just after the reign of Suleyman the Magnificent, when the Turkish Empire stretched in a huge arc from Hungary through the entire Middle East to North Africa. Constantinople was one of the many classical cities now in the hands of this formidable power and, with its imperial and Christian past, the city still exercised a deep fascination on the European mind.

Constantinople by Buondelmonte, 1482. This is still the Christian city, for this is a manuscript copy of an earlier image, pre-dating the city's capture by the Turks in 1457. Hagia Sophia is prominent on the right, with the remains of the Roman hippodrome before it. The commercial quarter of Pera, with its Galata Tower, lies across the Golden Horn.
The British Library, Arundel MS 93, f.155

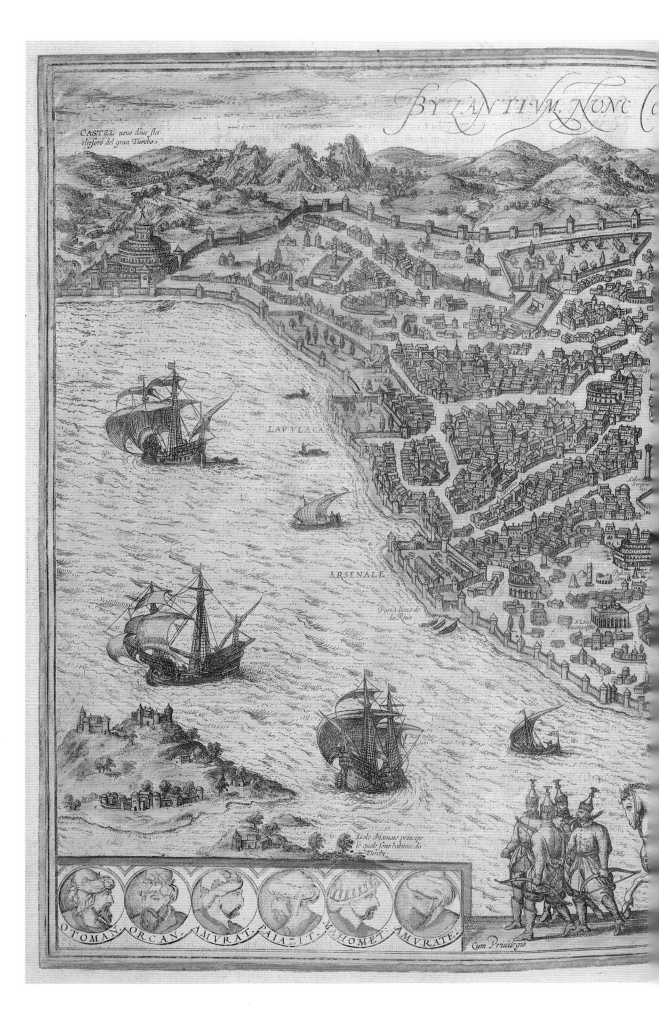

CASTEL nouo doue sta el tesoro del gran Turcho,

BYZANTIVM NONC C

Andrea

LAVVLACA

ARSENALE

Porta liona de la Riua

A sole chiamare principe le quale sono habitare da Turchi,

Cum Priuilegio

OTOMAN · ORCAN · AMVRAT · PAIAZIT · MAHOMET · AMVRATE ·

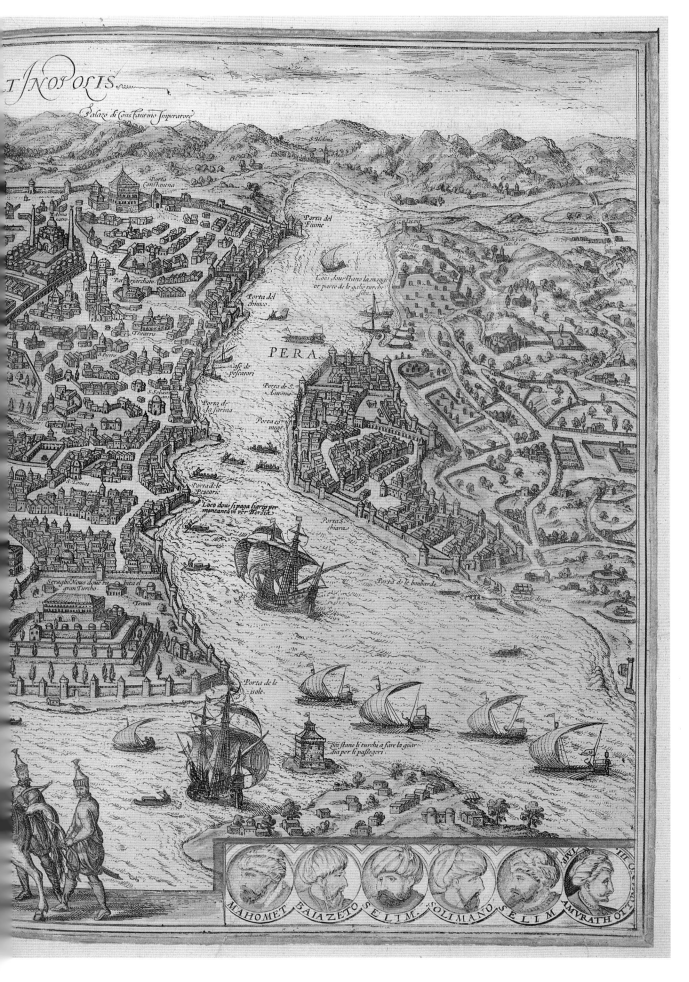

Constantinople
by Braun and
Hogenberg, 1572.
The Turkish city
is depicted here,
with the Sultan's
palace dominant
in the foreground,
mosques replacing
the churches, and a
gallery of portraits of
the Ottoman rulers.
The British Library,
Maps C.29.e.1 (51)

CUS

CVSCO, REGNI PERV IN NOVO ORBE CAPVT.

Cuzco

Cuzco is the oldest continuously inhabited city in the Americas; no one knows exactly how old because the Inca people, whose capital it was, developed no system of writing, and therefore left no historical records. The pre-colonial history of Cuzco has been pieced together, partly by archaeology and partly from oral traditions collected by the few Spaniards sufficiently interested to record them, such as Bernabe Cobo. The city lies at the confluence of two rivers, the Tullumayo and the Huatanay, surrounded by high mountain peaks, and more than two miles above sea level in the thin cold air of the Andes. Its wide plazas and baroque churches are typical of Spanish colonial architecture, while many narrow, stone-flagged side-streets still survive from the Inca period. Among its people, too, a mixed inheritance is still present, with Quechua, the Inca language, still flourishing alongside Spanish.

The earliest precise date in Cuzco's history is the fateful year of 1533, when a small but ruthless Spanish force, led by Francisco Pizarro, entered the city. They found a community of around 200,000 people, which formed the capital of an empire of perhaps ten million, stretching from northern Ecuador to southern Chile. The Spaniards learnt that the name Cuzco meant 'the navel of the world'. The heart of this capital was the Coricancha Temple, the cultic centre devoted to the sun-god, but containing also shrines to the gods of the moon, stars, thunder and rainbow, as well as a number of government offices and the central observatory from which astronomers regulated the calendar and the festivals of the religious year. One mile north of Coricancha, on a hilltop overlooking the city, was the massive fortress of Sacsahuaman, one of the most colossal structures of pre-Columbian America. Both temple and fortress probably date from 1440 to 1480, and were the work of the Inca ruler Pachacuti. These huge structures were built without the use of the wheel or of iron tools. Another impressive feat of engineering was the diversion of the Tullumayo River to provide water channels throughout the city. Two generations after Pachacuti's reign, Inca society was eaten away by civil war and fatally weakened just on the eve of the Spaniards' arrival.

In Spanish eyes the great lure of Cuzco was the mass of gold and silver adorning its temples and palaces. Within a year they had shattered the Inca forces, murdered the last king, Atahualpa, and looted the city. Two years later however, in 1536, Cuzco became the focal point of a last desperate campaign of Inca resistance, which proved futile and which led to the destruction of much of the city, including the Coricancha Temple and the Sacsahuaman fortress. The Spanish moved their capital city to Lima on the coast, while Cuzco was rebuilt, replacing temples with churches but preserving the lines of Pachacuti's streets. A violent earthquake in 1650 necessitated much rebuilding and decoration in the emerging 'Cuzco style' of Spanish baroque. Cuzco never lost its place as the second city of Peru, but it acquired a new importance after the year 1911, when Hiram Bingham discovered the unique Inca settlement of Machu Picchu, hidden for centuries in mountains 70 miles (113 km) to the north-west. Its existence had apparently remained unknown to the Spanish and, with its discovery, Cuzco soon became Peru's major centre of tourism and of the study of Andean civilisation. The destruction of the city and of the Inca culture, so ruthless and so complete as it must have seemed in 1536, has proved after all not to be permanent.

Cuzco by Braun and Hogenberg, 1572. This is a version of the only image of the Inca city of Cuzco available in Europe in the sixteenth century. The Inca king borne on his litter has been added to give dramatic interest. We have no way of knowing how accurate the plan is, but the walls and gates look distinctly European. The Temple of the Sun is on the left, but the fortress of Sacsahuaman is not in view.
The British Library, Maps C29.e.1 (1)

Delhi, 1851. A view of the old city, seen across the Yamuna
Bridge, relatively small and compact, and dominated by
the Red Fort. It is not yet the capital of the Raj, and the
events of the Mutiny are some years in the future.

Delhi

Located at a focal point of northern India, midway between the Himalayan kingdoms, the Indus Valley and the plain of the Ganges, Delhi has been the site of a succession of cities for almost 3,000 years. The exact location has shifted a number of times, and of the earliest Hindu settlements there is now almost no trace; Delhi's history is overwhelmingly Islamic and then British. The importance of the Asoka Pillar lies in its isolated testimony to the rule of the Buddhist emperor of the region in the third century BC. During the twelfth century AD, India's patchwork of Hindu states was progressively attacked by Muslim armies from Afghanistan, and in 1192 Delhi itself fell to the invaders. The earliest monument of Islamic rule is the Qutb Minar south of the present city, dated around 1206, which, according to its inscriptions, was intended to 'cast the shadow of Allah on east and west'. The first mosques were built from the stones of shattered temples. The Delhi Sultanate was established and Mongol pressure on the Islamic heartlands of the Middle East enhanced Delhi's status as a centre of Islamic culture. When Baghdad fell to the Mongols in 1298, refugee scholars and artists flocked to Delhi. But a century later disaster overwhelmed the city when the Mongol hordes of Timur (Tamerlane) swept through northern India, destroyed Delhi and slaughtered its people. Delhi took almost a century to recover, as the Sultans moved south to Agra, away from the scene of devastation.

Delhi's re-emergence into history really began in 1526 when Babur, an Afghan warlord, defeated the last sultan in the great battle of Panipat, 50 miles (80 km) north of the city, and founded the Mughal Empire. Over the next century the Mughal rulers Humayun, Akbar, Shah Jahan and Aurangzeb developed Delhi and Agra as twin capitals. The characteristic Mughal style of art and architecture flourished in both cities, in the Red Fort, the Jami Masjid Mosque and the many royal tombs. Although Shah Jahan is probably best remembered as the creator of the Taj Mahal in Agra, it was he who finally shifted the seat of government back to Delhi, which became known as Shajahanabad and is now part of Old Delhi. The British grasp on Mughal India was virtually completed in 1772, when the East India Company established its residents in Delhi in order to exercise their decisive influence over the Mughal rulers, nominally still their overlords but now

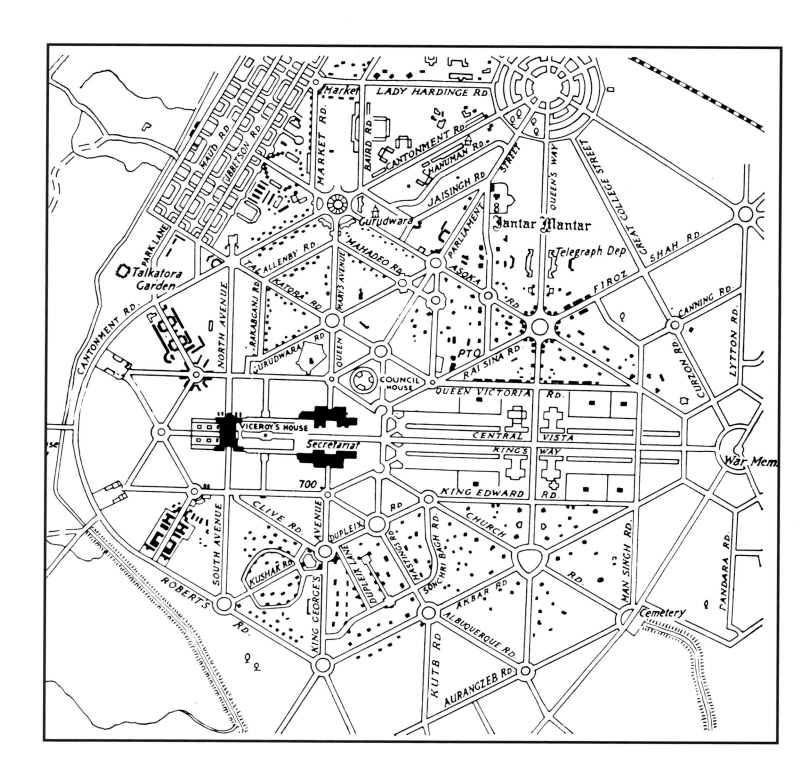

New Delhi, designed by Lutyens and Baker between
1911 and 1916: a grandiose, formal plan suited to an
imperial capital, and unrelated to the haphazard old
town to the north.

The British Library, YH1989a133

their puppets. Delhi was the greatest rebel conquest during the Indian Mutiny of 1857, and hundreds were massacred when the Red Fort fell. The aftermath of the Mutiny saw savage reprisals and the end of the Mughal dynasty, for the last king, Bahadur Shah II, was tried for complicity in the Mutiny and exiled.

Calcutta, original headquarters of the East India Company, had assumed the role of capital of British India, yet in 1902 it was Delhi which Lord Curzon, the Viceroy, chose to be the scene of the great durbar to celebrate the accession of King Edward VII. Nine years later, when the king visited India in person, he announced that Delhi was to be rebuilt as the new capital. There were practical merits in this move, for Delhi was more central than Calcutta and better served by the railway; but there is little doubt that the real reason was the desire of the British to proclaim themselves the true heirs of the Mughal Empire. The site chosen for New Delhi covered some 10 square miles (26 square km) and was centred on Raisina Hill, south of the old city. Two leading British architects, Edwin Lutyens and Herbert Baker, were commissioned

to design a spacious, triumphalist capital. They embraced this challenge to create a style of architecture that was, in Baker's words, 'not Roman, not British, not Indian but Imperial'. The symbolic centre was to be a two-mile-long grand vista, running east from the viceroy's palace, itself a modern Versailles. The administrative buildings flanked this vista, while diagonal avenues radiated outwards to the residential and business districts, the latter centred on the great wheel of Connaught Place. (The parallels with the plan of Washington are very striking.) As construction proceeded during 1912 to 1916, Lutyens realised with horror that something had gone wrong with his design: the gradient of the grand vista had changed, with the effect that the intended view of the viceroy's palace was obscured. Lutyens's horrified protests have become part of Delhi's history, yet to a large extent he and Baker had succeeded admirably and New Delhi is a functional and aesthetic success. It has grown with the nation; the Council House has become the Parliament, and its status as the rightful capital of India has never been questioned.

Delhi's Red Fort, built by Shah Jehan in the mid seventeenth century, drawn and lithographed by James Duffield Harding in 1847, after Charles Stewart Hardinge.
The British Library, Cup.652.m.34

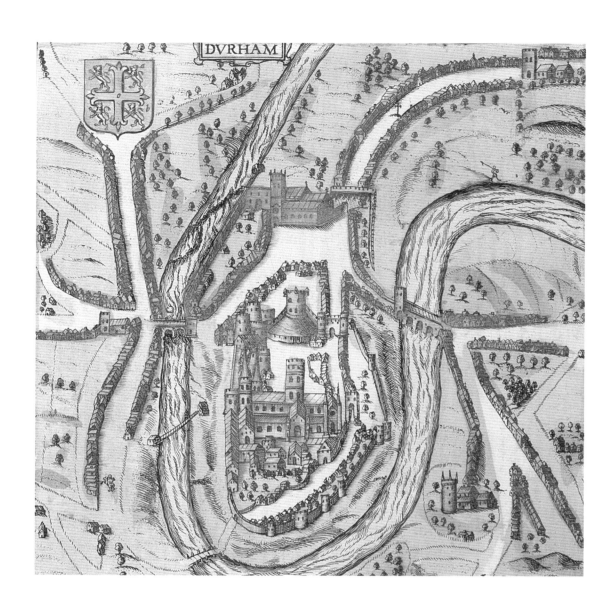

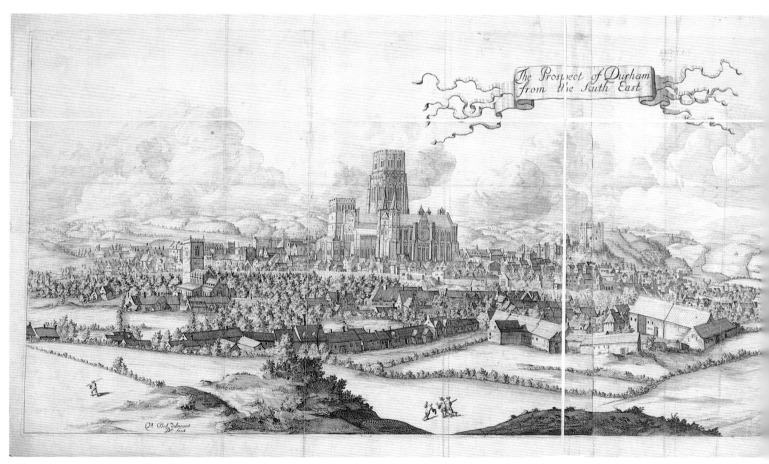

Durham

If ever a site invited the building of a city, it is surely that of Durham - a ridge of rock, half a mile in length, that rises high above a tightly folded meander in the River Wear, the site's layout strikingly resembles that of Bern in Switzerland. Traces of pre-historic settlement have been found below the town, but mysteriously it was somehow overlooked by the Romans so that its history really begins in the year 995, when a group of monks, driven from Lindisfarne by the Viking raiders, built here a shrine for the relics of St Cuthbert. In the years following the Norman conquest, a monastery, a cathedral and a castle - the three great emblems of medieval culture - were erected on the hill. The peninsula was encircled with high stone ramparts, a circular castle was reared at the apex of the site, and on the highest point of all, approached through narrow winding streets, the cathedral was built in the massive Romanesque style. This architectural complex functioned as the visible symbols of the new Latin civilisation which now towered over its Nordic and Saxon subjects.

From the first, Durham's bishop fulfilled a dual role, temporal as well as spiritual, a prince-bishop, sovereign in his own 'Palatinate', dispensing justice, minting coinage and leading armies. The principality was a buffer between England and Scotland. One of the most celebrated medieval bishops, Anthony Beck, personally led almost 2,000 men when assisting Edward I's invasion of Scotland in 1296, and was repaid with the honorary title of King of the Isle of Man. Some of these secular privileges survived into the nineteenth century. The shrine of St Cuthbert was a place of pilgrimage until it was suppressed at the Reformation, while during the Commonwealth period, the Palatinate was temporarily abolished, and the cathedral was used by the Puritans as a military prison in which their enemies were confined.

For centuries the topography of the site strictly limited Durham's growth, although it was natural that outlying communities should spring up around the city's two bridges - Elvet in the east and Framwellgate in the west. When the railway line was taken past the city to the west, the Framwellgate area developed around it into a virtual suburb. However, the overwhelming fact of Durham's history in the nineteenth century was a negative one: the city remained untouched by the industrial revolution. Coal-mining and iron-working developed in the surrounding towns and villages, but not in the city itself, or even its immediate vicinity. It never became a centre of trade or industry, but by an accident of history remained an ecclesiastical and administrative centre, a kind of northern Salisbury or Winchester; its architectural splendour was preserved, and it was rapidly overtaken in size by a great many industrial towns in the region which it governed. When the Palatinate was finally abolished in the reforming era of the 1830s, the Castle, the principal residence of the Bishop, was given to become the centre of the newly founded university, the first to be established in England after Oxford and Cambridge. (Strangely perhaps, it was a Durham graduate, Edward Bradley, who wrote the classic comedy of undergraduate life in Victorian Oxford, *The Adventures of Mr. Verdant Green*, 1853-57.) The growth of the university in the twentieth century was responsible for the city's spreading to the south for the first time, beyond the loop of the river, although the historic character of the city centre, secure on its peninsula, remains unchanged.

Left: Durham, by Speed, 1611. No city centre could show greater historical continuity than Durham, and a glance at this Tudor map reveals why: the castle and cathedral, islanded in the bend of the River Wear, have created a pattern that simply cannot be changed, whatever developments may appear in the outer districts.
The British Library, Maps C.7.c.20

Below: Durham, a delightful late eighteenth-century view from the south-east, in which the cathedral dominates, but in which the artist has contrived to hide the river entirely.
The British Library, Maps 2265 (1)

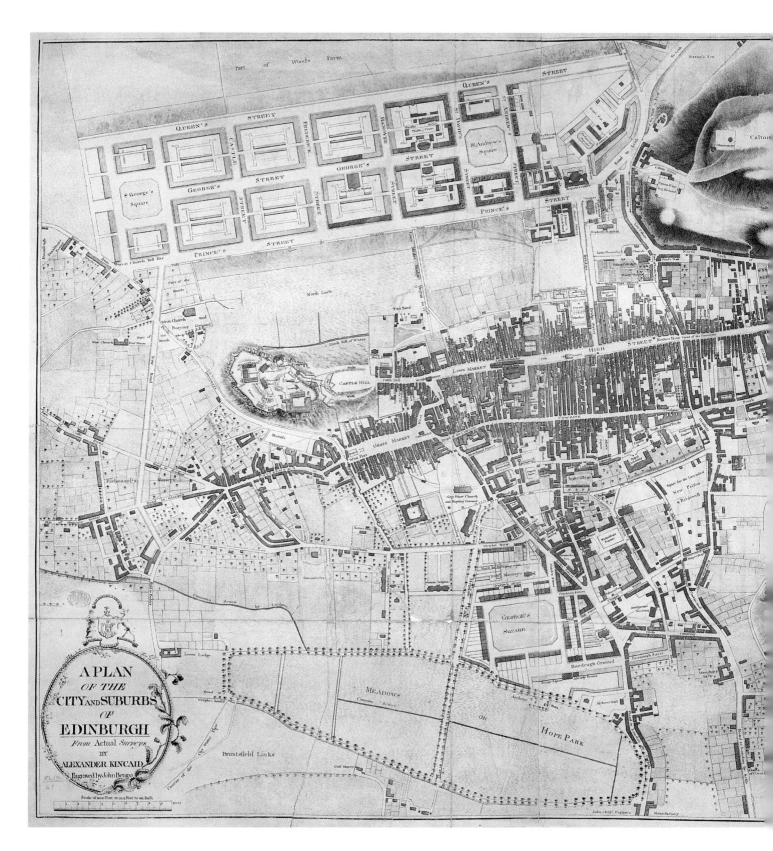

Edinburgh by Kincaid, 1784. By the eighteenth
century the rising ground to the north of the
city is laid out in the elegant streets and
squares of Craig's New Town, linked to the
Old Town by the all-important bridge across
the marshy 'North Loch', where the railway
line now runs.

Edinburgh

The city of Edinburgh possesses a theatricality that is unique among British cities. This is partly physical, arising from its setting among crags and hills, its dark stone buildings and the open space beneath the castle which seems so vast for a city centre. The writer Robert Louis Stevenson called Edinburgh 'a dream in rock and masonry'. But the sense of drama resides partly in our minds too, for we are always conscious of the city's past - not merely the pageant-past of the Stuart monarchs, the Reformation and the Jacobite Rebellion, but also the past of the common people, the city of conviviality, drunkenness, crime and dark alleys, the Edinburgh of Burke and Hare, or Jekyll and Hyde, the city in which Conan Doyle grew up. With this physical setting and this past, and with the people of Edinburgh - by turns warm, articulate, riotous, canny, capricious - it is evident that no other city could possibly fulfil the role of Scotland's capital.

Until the 1760s Edinburgh meant the 'Old Town', stretching along the ridge from the Castle Rock down to the Palace of Holyrood. Here the seats of national power - castle, cathedral, parliament and high courts - were surrounded by the tall, narrow tenements which housed rich and poor alike. This unlikely setting was transformed into the 'Athens of the North' by the presence in the eighteenth century of a group of leading thinkers - Hume the philosopher, Adam Smith the economist, Joseph Black the scientist, James Hutton the geologist, and many others. Publishing and journalism flourished, and *The Edinburgh Review* and *Blackwood's Magazine* dominated literary life throughout the nineteenth century, while the university rivalled any in Europe.

The city expanded steadily southwards into the meadows where the university and hospitals were built, but to the north the way was blocked by a marshy moraine known the 'North Loch'. The great event in Edinburgh's architectural history was the draining of this loch, and the bridging of the resulting valley in 1772. This opened the way for the New Town to be built, and the architect James Craig designed wide geometric streets and elegant Georgian houses to cap the low hill to the north. The valley provided the ideal route to the city centre for the railway when it arrived in the 1840s, while Princes Street Gardens forms a perfect boundary between the old and new towns.

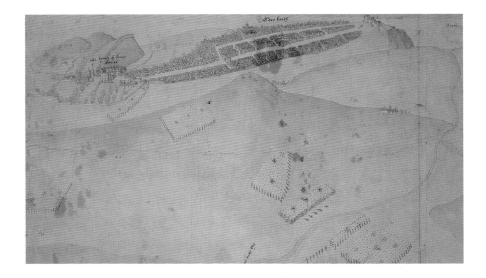

Edinburgh, 1544. This manuscript is the first known view of the city, drawn by English soldiers who were besieging the city. The Old Town is seen from the north, strung out along the Royal Mile from the Castle down to Holyrood - '*the King of Skotes palas*'. The English army is encamped to the north, where the New Town will later be built.
The British Library, Cotton MS Augustus I ii 56

Florence, c.1485. Francesco Rosselli's large woodcut
of Florence is the earliest detailed and authentic view
of any European city. The lock in the top left corner is
an enigmatic symbol that has never been explained.
In the lower right corner is an intriguing self-portrait
of the artist at work drawing this panorama.
Staatliche Museen, Berlin

Florence

Florence is identified not with a continuous past, not with centuries of glorious history, but with a single period of perhaps eighty or ninety years when the city experienced a unique outburst of creative energy. All histories of the Renaissance begin in Florence, but the reasons for this can never be finally explained. An unimportant Roman town, none of whose classical past remains, it emerged from centuries of obscurity into the Middle Ages with one unusual characteristic: it was a commune, a kind of republic, which was not dominated by prince or pope. This was not a democracy, but power in Florence was in the hands of a cultivated bourgeois elite, whose twin aims were to acquire wealth and to spend it splendidly. This did not mean that the city was peaceful, on the contrary it was wracked with feuding and factionalism - with the most famous exile from these conflicts being Dante. But Florence in the late Middle Ages was somehow intellectually an open city in a way that Rome or Naples or Venice was not, and art flourished there outside the ducal palace or the cathedral. Its cultural rise occurred in precisely the years when Rome was in deepest eclipse, a fact much vaunted by Florentine writers who saw their city as inaugurating a new age as glorious as the classical age itself. The idea of the Renaissance is not an invention of modern historians: it was used by contemporaries to describe the rebirth which ended the post-Roman darkness. 'It is undoubtedly a golden age', wrote one Florentine, 'which has restored to the light the liberal arts that had almost been destroyed: grammar, poetry, eloquence, painting, sculpture, architecture and music - and all in Florence.'

The list of individuals who achieved this transformation are the giants of the Renaissance: Dante and Boccaccio had already fashioned the Tuscan dialect into the literary language of Italy; Masaccio, Piero della Francesca and Botticelli brought painting into the new age; Donatello, Brunelleschi and Michelangelo recovered the Greek ideals in sculpture and architecture. These are the artists who shaped the fabric of Florence, in which the modern visitor finds the spirit of the Renaissance. Their self-conscious cult of beauty created buildings which should be treasures, whose purpose was in turn to house more treasures. The remarkably unchanged skyline of the city when seen from the hills - the cathedral and the Palazzo are still the tallest buildings - sustains the conviction that we are still able to walk the streets of the city where the Renaissance was born.

The wealth of Florence grew up on the twin foundations of cloth and banking. Wool from as far afield as England and Flanders was brought into the city and worked into bolts of cloth by thousands of weavers, while much smaller quantities of silk from China were also processed. Modest fortunes made in this way were diverted into banking, until the leading families in that field were known throughout Europe, not only the Medici but also the Bardi, the Strozzi, the Pazzi and the Pitti. The Medici owed their pre-eminence to their role as bankers to the papacy. It was not until 1569 that the title of Duke of Florence was taken by Cosimo; until that time the Medici had ruled simply as first among equals, in which time they had married into the royal houses of Europe and provided the church with three popes (including Leo X, the most worldly and extravagant of the Renaissance popes). This wealth was poured into the patronage of the arts, with historic results. But there was a religious reaction against the cult of beauty, and the most dramatic years in the history of Florence were the 1490s, when the charismatic preacher Savonarola thundered against luxury and self-indulgence, and drove the Medici into exile. For a brief period Savonarola was virtually theocratic ruler of the city, until his fall and terrible death by fire in the Piazza della Signoria. The republic was finally extinguished when the city fell to a French army in 1532. The Medici were restored to a more autocratic form of power, and the focus of power and creativity in Renaissance Italy shifted to Rome. Florence today is a monument to the artistic energy which surged through the city for those few decades late in the fifteenth century. It is a living city too, but it is, above all, the past which lives there.

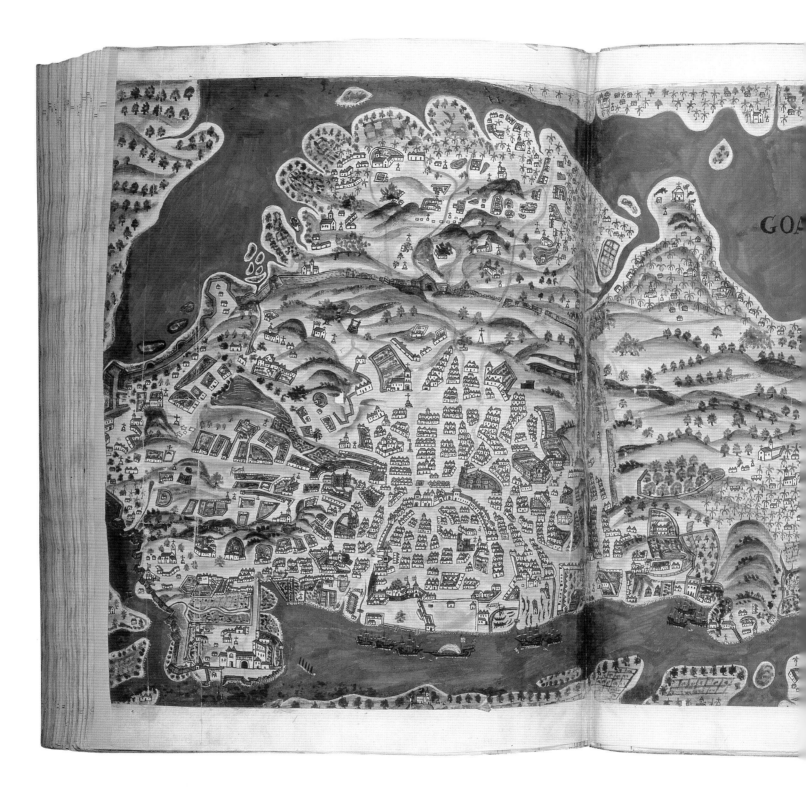

Goa, a hand-drawn view of 1646 by Pedro de Resende, illustrating a history of the Portuguese trading empire in the east. The island has been drawn with north at the bottom, in order to bring the city itself into the foreground. The naïve use of multiple perspectives in the drawing of the buildings makes one wonder if this colourful plan was based on native Indian sketches.
The British Library, Sloane MS 197, ff.247v–8

Goa

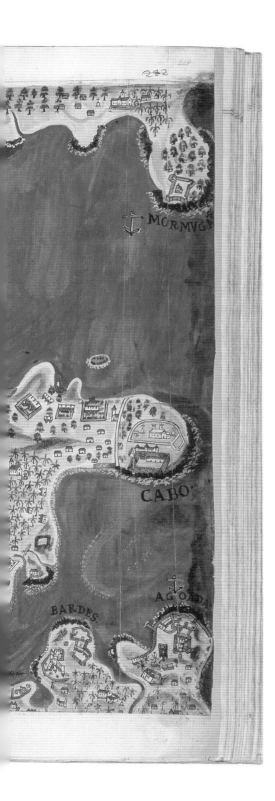

The old city of Goa is just as much a monument to a vanished civilisation as Pompeii or Teotihuacan, yet for a hundred years between 1530 and 1630 it was a city to rival any in Europe, a centre of trade, ecclesiastical power and baroque art. A Muslim city with good harbour facilities and situated on a peninsula between two river estuaries, it was quickly seen by the Portuguese commander Afonso de Albuquerque to be the ideal capital for his country's aggressive campaign to dominate commerce in the Indian Ocean. Following da Gama's historic first voyage to India in 1498, the Portuguese aims were to secure an absolute monopoly of the spice trade, and to attack Muslim interests wherever possible. Albuquerque arrived in 1510 with an invasion fleet to take the peninsula. After a prolonged campaign, the Portuguese succeeded, massacred the Muslim defenders, and proceeded to rebuild Goa as the capital of their far-flung trading empire, the naval base of their powerful fleet, and the essential port of call to the Spice Islands.

The adjacent territory, some thousand square miles, became a Portuguese colony supplying the city with food. Portuguese men married Indian women and engendered a *mestizo* population, with the characteristic names which still survive – Fernandes, de Sousa, Gomes and so on. Churches, monasteries, hospitals, an arsenal and a new harbour were built, and by the 1590s the population reached around 100,000. The elite of Goa lived in great style, their indolence sustained by a brutal traffic in slaves brought from Africa and the East Indies. Portuguese imperialism was commercial, racial and religious – Goa became the centre for Counter-Reformation missions to Asia. The great Jesuit, St Francis Xavier, worked here for some years and after his death in 1552 his body was entombed in Bom Jesu Church, where it was for many years reputed to be miraculously preserved from corruption. Various macabre stories were connected with it, and today the crumbling and mummified corpse is still occasionally exhibited. The Portuguese flocked in their thousands to Goa in search of wealth or military glory, but among those who found neither was Portugal's national poet Luis de Camoes, who was shipwrecked, half-blinded and ruined by his Indian voyage; his statue stands before the Se Cathedral in Goa, the largest Christian church in Asia.

Goa's prosperity was short-lived however: after the Portuguese throne was annexed by Spain in 1580, Spain's European enemies attacked her overseas empire. The Dutch in particular seized her Asian territories and decimated her maritime trade, so that by 1650 Goa remained as the isolated capital of a vanished empire. At the same time cholera had become endemic in the city, tempting more and more of the Portuguese to move to other towns in the colony. After decades of decline, and long after the Viceroy had actually departed, the Portuguese officially moved the capital south to Panaji in 1760. The old city was left in the hands of priests and nuns, who strove to maintain the church buildings while the population dwindled to a few thousand. When the great traveller, Sir Richard Burton, visited Goa in 1850 he described it as a ghost town: the vast cathedral which would have graced any capital in Europe attracted twenty or thirty worshippers, and its walls were grappled by poisonous plants and thorny trees, like an ancient temple half lost in the encroaching jungle. In recent years the old city has been better preserved, but the neglect of almost two centuries can never be erased. After Indian independence, pressure was applied to integrate Goa into the new nation. Portugal resisted stubbornly for years, until the exasperated Indian government sent its army over the border in 1961, and four-and-a-half centuries of colonial history were brought to an end.

FRETI DANICI OR SVNDT ACCVRATISS DELINEATIO.

HELSEBVRGVM

Pharus

NO BI LISSI MVM

CORONEBVRGV

HELSCHENO

Interior Arcis magnificentia

Lundehoue

Helsingor

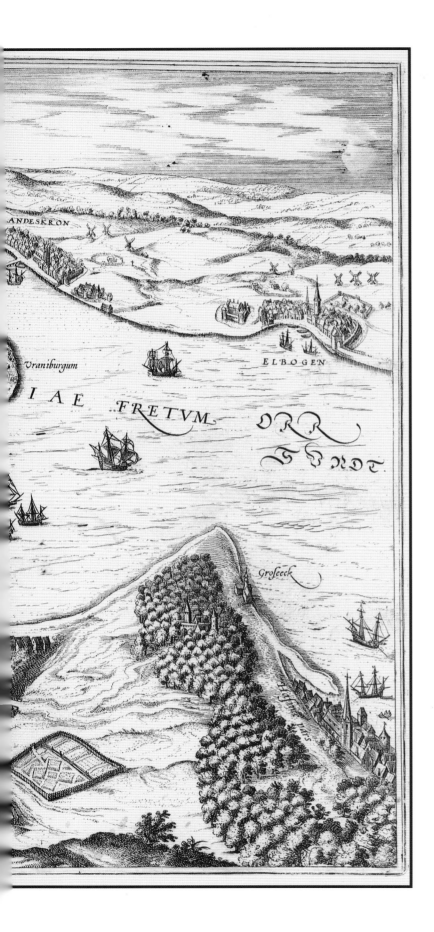

This view of the strait which separates Denmark from Sweden contains two features of exceptional interest – one literary, the other scientific. The first is the fortress of Kronborg, which dominates the small Danish town of Helsingor, and which has long been identified as the setting of *Hamlet*. Hamlet's story appeared first in medieval Danish chronicles, but it was said to be pre-Christian so, like the story of King Lear in the English chronicles, it must be regarded as legendary rather than historical. The only tenuous link with Shakespeare is that the castle itself was built in the 1580s, and it is possible that Shakespeare may have read or heard some contemporary description of it. It was a royal residence of King Frederick II and his descendants, and it enforced the tolls levied for crossing the strait, upon which much of Helsingor's prosperity depended.

The second feature is the island visible in the strait, named as Hvena, but now called Ven. It was here that Tycho Brahe, the great Danish astronomer, built his historic observatory in 1576, under the patronage of Frederick II. For twenty years, Tycho and a team of assistants worked to build up the body of precise observations with which Tycho revolutionised western astronomy. The observatory was called *Uraniborg* – Castle of the Heavens – and it was one of the most important sites in the history of western science. Sadly, all traces of it vanished years ago. There is no historical link between Helsingor and Hamlet, yet such is the power of literature that a large tourist trade has sprung up around 'Hamlet's Elsinore', and Kronborg is visited as if it were a place of history instead of a place of imagination.

Helsingor, by de Wit, *c*.1660. Separated from Sweden by only three miles of sea, Helsingor had always been a ferry port of importance, but it is doubtful if it would have been included in seventeenth-century atlases of city maps had not the Kronborg castle-palace (the setting of *Hamlet*) been built in 1580.

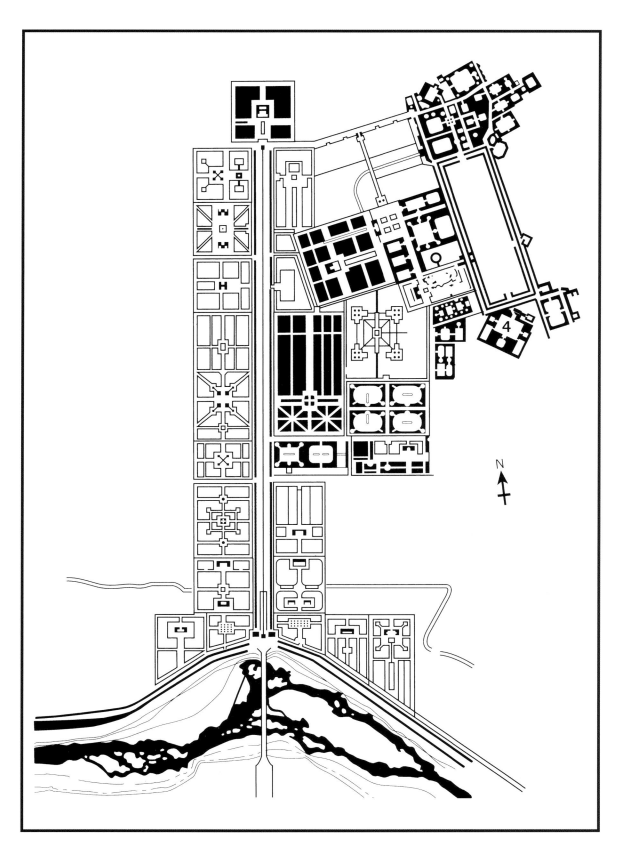

Above: Isfahan. No contemporary maps from the time of Isfahan's construction have survived. This modern plan from S. Cantacuzino and K. Browne's *Isfahan* (special issue of *Architectural Review*, London, 1976), conveys the grandeur of the two-mile-long avenue down to the River Zayandeh, the *Charar Bagh*, whose houses fronted the central thoroughfare, with its trees, canals, fountains and flowers. The mosques, bazaars and *maidan* are offset from the central axis.
The British Library, PP 1667ab

Right: Isfahan, the Shah Mosque, a masterpiece of Persian blue tile work, depicted in Coste's *Monuments de la Perse Moderne*, 1849.
The British Library, N. Tab 2000/2

Isfahan

The city, planned as a royal residence and showcase with palaces, avenues, vistas, gardens and places of worship, is far from being a European creation. Between 1598 and 1620, Shah Abbas I of Persia rebuilt the old city of Isfahan in the eastern foothills of the Zagros Mountains and made it the capital of the Safavid dynasty, whose splendour was captured in the proverb 'Isfahan is half the world'. Situated close to the Zayandeh River, the Shah chose a site two miles to the north to lay out a spacious new *maidan* – a grassy square or park – overlooked by the royal palace and by two exquisite mosques, their domes of blue and gold rising against the nearby mountains. Extensive covered bazaars bordered the *maidan*, bazaars which were high and spacious, almost resembling Italian gallerias. The *maidan* itself was the setting of ceremonies, parades, and games such as polo. This area of the city was connected to the river by a magnificent avenue fully two miles long, adorned with fountains, pools and trees, and flanked by the houses of the nobility. Some of these houses have been recently restored and are of great beauty. This avenue terminated in a marvellous bridge of no fewer than thirty-three arches, linking the new Isfahan to the older city. Some years later a second bridge was built downstream, which functioned also as a dam; it featured elaborate terraces, from which the turquoise mountain water could be admired. No comparable city was to be seen anywhere, and the European diplomats who came to the Shah's court compared it to an oriental paradise.

Europeans were eager to befriend the Persians, for the sake of the trade in silks, jewels and carpets, and in order to outflank the Ottoman Turks, the common enemy of Europe and of Persia. A number of visitors have left descriptions of the brilliance of Isfahan, including the two English brothers, John and Robert Sherley; the latter married a Persian princess, and served the Shah for many years, rather as Marco Polo had served Kublai Khan. Abbas's court represented the high point in the history of Persian decorative arts, but the Shah was still capable of extreme cruelty, torturing and executing members of his own family if he felt they threatened him.

By the close of the seventeenth century, Savafid rule in Persia was in decline, and in 1722 Isfahan was overrun and partly destroyed by Afghan invaders. Its fortunes revived in the twentieth century, and the newer town has expanded greatly to the north-west. Isfahan has recently been recognised as one of the most graceful and important achievements in urban planning anywhere in the world – displaying the kind of deliberate nobility that could perhaps be created only by a single-minded, autocratic ruler.

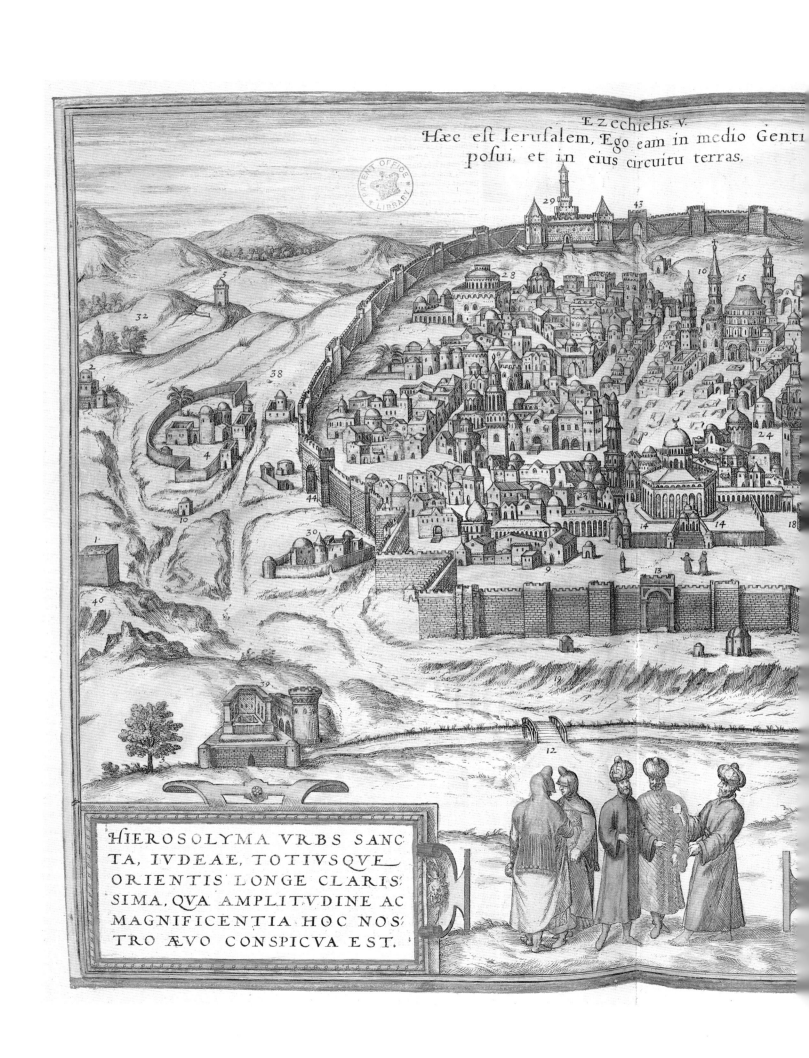

Ezechielis. v.
Hæc est Ierusalem, Ego eam in medio Genti
posui, et in eius circuitu terras.

HIEROSOLYMA VRBS SANC
TA, IVDEAE, TOTIVSQVE
ORIENTIS LONGE CLARIS:
SIMA, QVA AMPLITVDINE AC
MAGNIFICENTIA HOC NOS:
TRO ÆVO CONSPICVA EST.

Jerusalem

Three of the world's great religions venerate Jerusalem as the unique setting of man's historical encounter with God. In consequence, it bears such a weight of history and aspiration that it is less a city than an idea, a symbol, whose power surpasses even that of Rome. The word Jerusalem means, approximately, 'place of the God of peace', yet surely no city on earth has been more bitterly fought over and pillaged, conquered and reconquered. Egyptians, Babylonians, Persians, Greeks, Romans, Arabs, Crusaders and Turks have all controlled or savaged the city, and only for a fraction of its history has it been ruled by the Jews who claim it as their spiritual home.

Inhabited and fortified for more than four thousand years, Jerusalem's real history begins around 1000 BC with its adoption by King David as the cultic centre of the twelve tribes of Israel, home to the Ark of the Covenant, which had formerly been moved from site to site in the mountains and deserts. Solomon, David's son, built the first Temple around 950 BC. The royal and religious centre was built across two hills which sloped towards the south, and part of the intervening valley. It was defended by massive walls, and enclosed springs and pools. To the east across the Kidron Valley lay a third hill, the Mount of Olives. From an early date it was known also as Zion, a word of uncertain meaning. This city and its Temple were destroyed in 586 BC by the Babylonians, who carried much of the population into exile. Sixty years later, the Persian conquest of Babylon freed the Israelites to return and rebuild their Temple. The campaigns of Alexander the Great brought Jerusalem into the orbit of Hellenistic politics, as a territory of the Seleucid kingdom. It was the desecration of the Temple by King Antiochus Epiphanes which sparked the Maccabean revolt of 167 BC, which won a century of Jewish freedom. This ended in 63 BC when the Romans asserted authority over Palestine. Herod, appointed by the Romans as governor and king, besieged and captured Jerusalem in 37 BC. He went on to build theatres, hippodromes and palaces, and also to rebuild the Temple: the Wailing Wall is the sole surviving part of this structure. Herod's city was the Jerusalem which Jesus knew, but one of the great enigmas of the city's ancient topography is that the site of the crucifixion - Golgotha, the place of the skull - cannot be identified. Two Jewish revolts against Rome, in 66–70 AD, and in 132–135 AD, resulted in almost total destruction, and determined the Emperor Hadrian to establish there a Roman city, Aelia Capitolina, and to ban the Jews from the site.

Jerusalem's fortunes changed dramatically with the conversion of Constantine to Christianity in 312 AD. Pilgrims began to arrive, among them the Emperor's mother, Helena, intent on locating the place of the crucifixion and entombment of Christ. A site was indeed found, outside the walls of Herod's city, beneath which were certain tombs and, so legend relates, a wooden cross, which

Jerusalem, by Braun and Hogenberg, 1623. The view is from the east across the Kidron stream, through the Golden Gate to the Temple area. Like many such images of the city, different historical eras have been merged together, so that modern mosques and Christian sites are shown alongside 'Solomon's Temple'. The aim was obviously to place the great historic sites within the contemporary city of the seventeenth century.
The British Library, Maps C.29.e.1 (2)

Jerusalem seen across the Kidron Valley, one of the
superb hand-coloured drawings from David Roberts's
famous series of lithographs *Egypt and the Holy Land*, 1856;
at once highly accurate and highly romanticised, this was
the kind of image which inspired nineteenth-century
pilgrims to travel to Jerusalem.
The British Library, 1264.e.20

was hailed as the 'True Cross'. Many churches were built in the city, supreme among them the Church of the Holy Sepulchre, on that site. This golden age – for Christians at least – lasted for three centuries, until a Persian army, in the course of its war against Byzantium in 614, ravaged the area and destroyed much of Jerusalem. Just twenty-five years later, Muslim forces entered the city, and a new phase in Jerusalem's history began. At first tolerant of the Christian presence, later Caliphs of the eleventh century first destroyed the Christian shrines and then banned pilgrims from the city, thus provoking the crusades. In 690 the Dome of the Rock had been built on the Temple Mount, on a site associated first with Abraham's sacrifice of Isaac, but still more importantly with Muhammad's ascent to heaven. The crusaders' goal of recapturing Jerusalem was won in 1099: churches were rededicated – the present church of the Holy Sepulchre dates from this period – while Jews and Muslims were expelled. Saladin's army retook the city in 1187 and, with brief intervals, it remained under Muslim control until 1917 when Turkish power in the region collapsed. In the sixteenth century the Ottoman Turks built the walls which still define the Old City, and for three centuries Jerusalem did not grow, confined within these boundaries.

These were all external events; but the salient fact of Jerusalem's history is its symbolic importance to the religious imagination over more than two thousand years. The root of this symbolism is the perception of this city as the dwelling-place of God. This originated with David's placing of the Ark of the Covenant here, and for Christians this was extended and deepened when Christ chose Jerusalem as the scene of his ministry and death. But this holy site had been repeatedly destroyed by human folly and violence, and thus was born the perennial dream of rebuilding Jerusalem, of restoring sanctity to the world of man. Sometimes this new creation was to be a city of bricks and stones, but sometimes it was conceived as a heavenly city, whose appearance would signal the end of time. Medieval geographers expressed the centrality of Jerusalem to their faith by drawing maps in which Jerusalem lay at the precise centre of the world, thus giving a literal interpretation to Ezekiel's text that Jerusalem lay 'in the midst of the nations'. The impulse to build an ideal city, a theocracy, a new Zion, has possessed many religious leaders, from sixteenth-century Germany to nineteenth-century America, and in a secularised form it was a guiding dream of the early socialists. The modern history of Jerusalem since 1917 has been shaped by the rise of Zionism and the building of the state of Israel, but the hatred and bloodshed which have filled that history show how hard it is to live in a city that is trapped rather than liberated by the religious visions of the past.

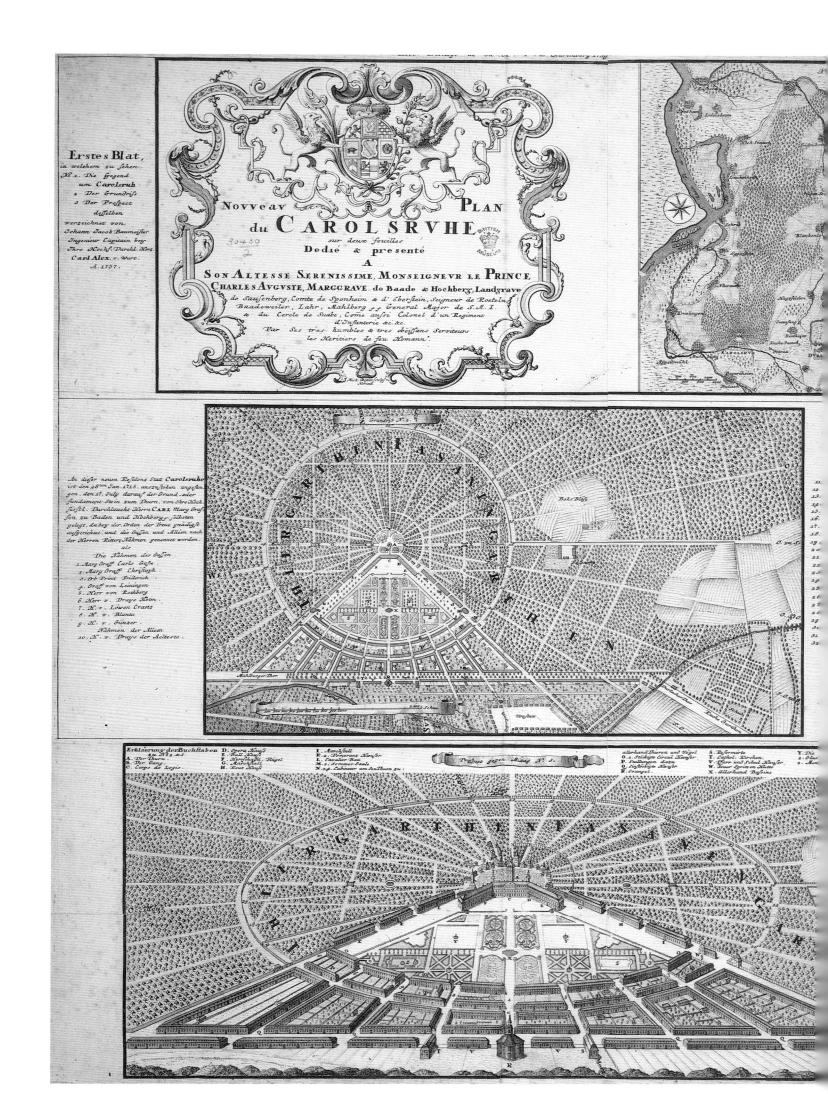

Karlsruhe

Founded in 1715 around a royal retreat-palace, Karlsruhe (Karl's *Ruhe* – Charles's 'rest' or 'peace') embodies a unique experiment in town planning. The palace of the Prince Karl Wilhelm, Elector of Baden, was placed in the centre of a circular clearing in a forest, and it was to form the nucleus of a small town. In a vast fan-shape, thirty-two avenues were made to radiate out from the palace, representing the thirty-two points of a compass. To the north of the palace, the avenues crossed a formal garden before leading into the forest, while the town was built on the southern side. Access roads crossed the radial avenues, and at these crossing-points the diagonal perspectives invariably led the eye up to the palace. Several decades were required to complete the design, and Balthasar Neumann, the doyen of German rococo architecture, was one of the leading planners. The market square with its equestrian statue of the founder was not completed until the end of the eighteenth century, and included arcades for stalls and shops, a curious mixture of ceremonial grandeur and everyday functionalism. It was as if a small market town should have been planted in the grounds of Versailles.

The political symbolism of the whole design is obvious: the palace and its ruler formed the centre of the world, around whom all else revolved. The principles behind the design were aesthetic and geometric: the radial city of Palmanova was one model, while the contemporary fashion for geometrical garden parterres was another. The city was consciously shaped as an intellectual exercise, appropriate to an age of absolutist rule. The motto of the founder was evidently, *L'etat c'est moi*, or more specifically *La ville c'est moi*.

The court of Karlsruhe was at various times host to Voltaire, Goethe and Gluck, and in 1806 the city became the capital of the Grand-Duchy of Baden. Karlsruhe has a secure niche in technological history: firstly, it was here in 1818 that Baron Drais of Sauerbrun invented the 'Draisienne', the hobby-horse on wheels which was the prototype of the bicycle, and from here the first cycling craze swept through Europe. Secondly, in the 1880s Heinrich Hertz first propagated electromagnetic waves – radio waves – in the laboratory of the technical school here. Throughout the nineteenth and twentieth centuries the modern town developed to the south and west, leaving the park and forest area to the north still in virtually their original form.

Opposite: Karlsruhe by Homann, 1739. Karlsruhe was designed like a cosmos in miniature, with the royal palace lying symbolically at its centre. From the palace, thirty-two avenues radiate outwards along all compass directions into the town and the countryside.
The British Library, Maps 30430 (2)

Above: Karlsruhe, a view of the central square from the palace towers, looking towards the surrounding forest.
The British Library, Maps 3.a.21

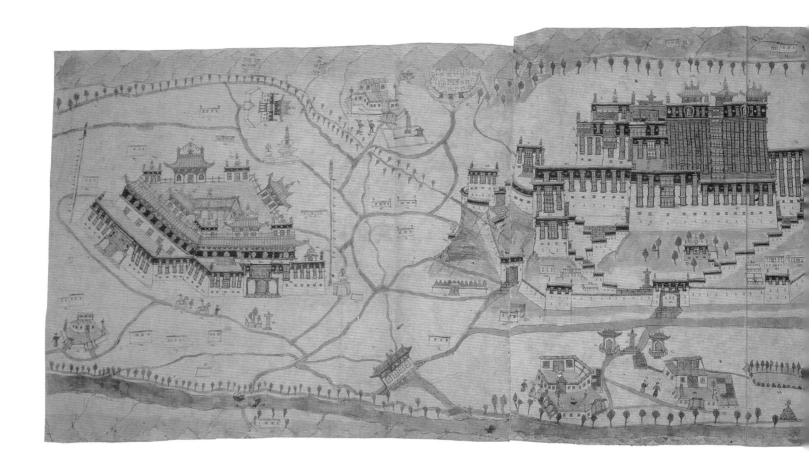

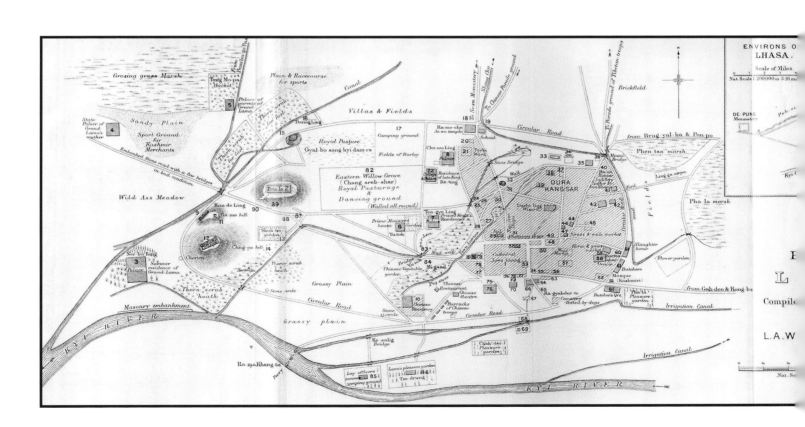

Lhasa

For two centuries or more the capital of Tibet has exercised a distinctive fascination on the European mind. Perhaps the most inaccessible city in the world, Lhasa lies 12,000 feet (3,600 metres) above sea level, isolated by the ramparts of the Himalayas to the south and elsewhere by a vast and desolate plateau. It was this very isolation, coupled with reports of the golden palaces and temples of a sphinx-like civilisation hiding itself from the outside world, which excited the imagination of adventurers, missionaries and imperialists alike. Their determination to break down the walls of Lhasa's isolation led eventually to tragedy for its people and its historic culture.

Lhasa was chosen as the capital city by Songsten Gampo, the seventh-century king who united Tibet as a nation and who established Buddhism as the state religion. It was Songsten Gampo who built the first Potala Palace on the Red Hill, although the structure which we now see is largely the creation of the seventeenth century. There is no doubt that the Potala – once the residence of the Dalai Lama and seat of Tibetan government – symbolises Lhasa to outsiders: the angular, towering palace clinging to the hillside high above the city must be one of the most unforgettable buildings in the world. Buddhism was apparently brought to Tibet by Songsten Gampo's Chinese wife, who founded the Jokhang Temple on Chokpori Hill, the other great spiritual centre of Lhasa, to which pilgrims have flocked for centuries.

Tibet's modern history has been largely shaped by its position as a border state between three empires – Russia, China and British India – for each has tried to prevent Tibet from falling under the power of the others. It was Chinese pressure which induced Tibet to close its borders to the outside world in 1792. This provoked a stream of individual Europeans to try throughout the nineteenth century to penetrate Lhasa's defences. All were unsuccessful, and the only authentic reports of Lhasa were brought back by the handful of spies sent out by the Survey of India, disguised as Buddhist pilgrims. This century of isolation was violently ended by the British military expedition in 1903-4 under Francis Younghusband. Its pretext was to secure a trade treaty between Tibet and British India; its true motive was to outflank Russian and Chinese designs on Tibet. The 'unveiling' of Lhasa aroused enormous public interest: books by eyewitnesses proliferated, and the first historic photographs of the city were published. In fact those who had seen the fabled city were unanimous in their disillusionment: the city which appeared so stunning from a distance was in reality squalid and filthy beyond belief. The streets were filled with stagnant pools, and dogs scavenged among the rubbish; the buildings were dark and fetid; the people were sullen and ragged; the city was unpaved, unsewered and unlit, like a survival from the Middle Ages.

Having stripped the mystique from Lhasa, the British, inexplicably, concluded a treaty with the Chinese which recognised their sovereignty over Tibet. The Chinese proceeded to occupy Lhasa for two years from 1910 to 1912, until the revolution within China itself forced their withdrawal. Tibet then enjoyed one generation of independence, until the second Chinese invasion of 1950, which has not yet ended. Half a century on, and the Chinese presence looks absolutely permanent. Lhasa's culture and its physical fabric are being systematically modernised and destroyed. The ruthless violence of the Cultural Revolution years, 1965-75, is over, and the destruction is now quieter, but still remorseless. This is undoubtedly being hastened by the growth of western tourism: what people hope to find in Lhasa is being destroyed by the thousands who go to see it. Western bars and discos, and the faceless sprawl of New Lhasa, are a sad fate for a city which only ever asked to be left alone.

Above: Lhasa around 1860. A magnificent anonymous painting of the Potala Palace and the Jokhang Temple, seen from the south across the Kyichu River, with the northern mountains visible in the background. The perspectives are not consistent throughout this view, but it is nevertheless a superb and detailed record of the old city which was hidden from western eyes for so long.
The British Library, Add Or 3013

Left: Lhasa, mapped by Waddell in 1904 and published in the *Geographical Journal* – one of several plans of the city to appear in the West after the return of members of the Younghusband expedition. The contrast which this dry, colourless plan makes with the vibrant Tibetan painting could hardly be stronger.
The British Library, Ac 6170/2

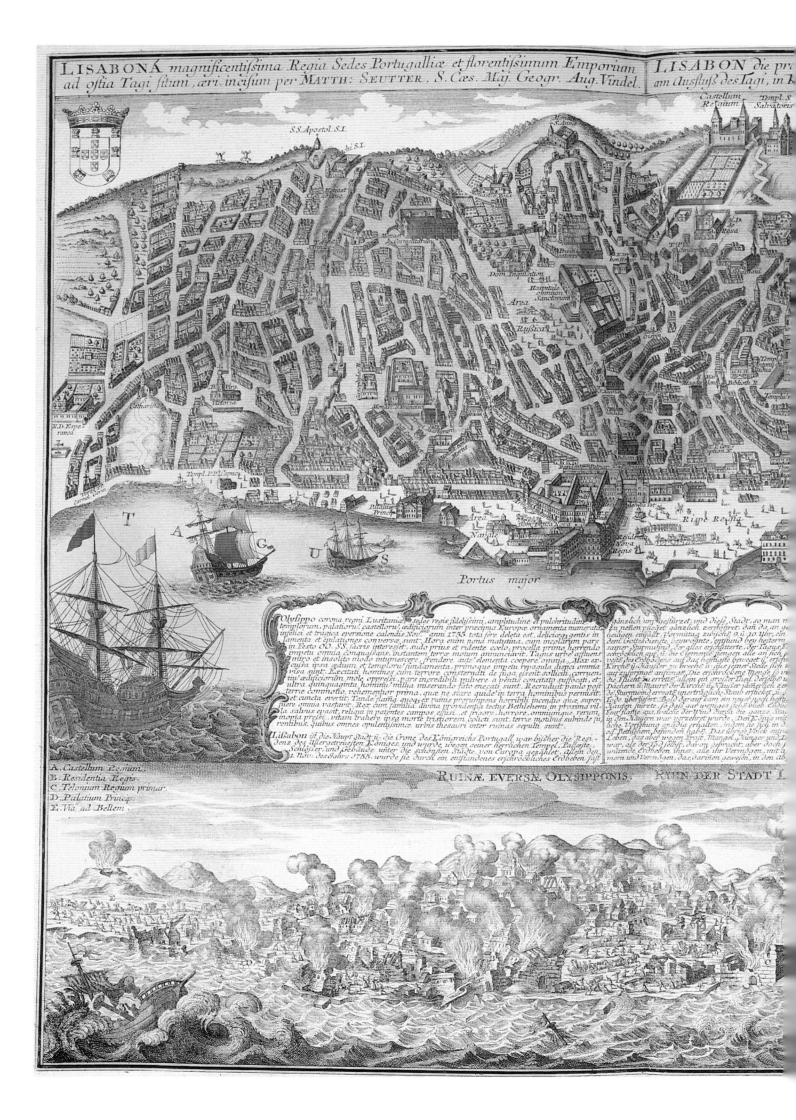

LISABONA *magnificentissima Regia Sedes Portugalliæ et florentissimum Emporium* LISABON *die pr..*
ad ostia Tagi situm, æri incisum per MATTH: SEUTTER, S. Cæs. Maj. Geogr. Aug. Vindel. *am Ausfluß des Tagi, in K..*

S.S. Apostol. S.I.

Castellum Regium   Templ. S. Salvatoris

T A G U S

Portus major

Riga Regia

Residentia Nova Regis

Olysippo corona regni Lusitaniæ, sedes regis fidelissimi, amplitudine et pulchritudine templorum, palatiorum, castellorum, ædificiorum inter præcipua Europæ ornamenta numerata, infelici et tragica eversione calendis Novembris anno 1755 tota fere deleta est, deliciæ gentis in lamenta et epulationes conversæ sunt. Horæ enim nona matutina, cum incolarum pars in festo OO. SS. sacris interesset, sudo prius et ridente cœlo, procella primo horrendo impetu omnia conquassans, instantem terræ motum annunciavit, Tagus urbe affluens mirio et insolito modo intumescere, frendere ante elementa cœpere omnia. Mox excussa ipsa ædium et templorum fundamenta, primæque ingeta repagula dejecta omnia visa sunt. Excitati homines cum terrore constantæ de sua ejectis sollicit. conuen tiu ædificiorum mole, oppressi, pars incredibili pulvere a ventis concitato suffocati, et ultra quinquaginta hominum milia miserando fato enecati sunt. Recruduit paulo post terre commotio, vehementior prima, quæ ne stare quidem in terra hominibus permisit, et cuncta evertit; Tandem flamma quoque ex ruinis prorumpens horribili incendio quæ, super fuere omnia vastauit. Rex cum familia, divina providentia tectus Bethlehemi in proxima villa salvus evasit, reliqui in patentes campos effusi, et frigore, horrore, omnimque rerum inopia pressi, vitam trahere ipsa morte tristiorem coacti sunt, terræ motibus subinde ju rentibus, quibus omnes opulentissimæ urbis thesauri inter ruinas sepulti sunt.

Lisabon ist die Haupt-Stadt u. die Crone des Königreichs Portugall, u war bißher die Residenz des allergetreuesten Königes, und wurde wegen seiner herrlichen Tempel, Pallaeste, Schlösser, und Gebäude unter die schönsten Städte von Europa gezählet. Allein den 1. Nov: des Jahrs 1755. wurde sie durch ein entstandenes erschröckliches Erdbeben fast

A. Castellum Regium.
B. Residentia Regis.
C. Telonium Regium primar.
D. Palatium Princip.
E. Via ad Bellem.

RUINÆ EVERSÆ OLYSIPPONIS.   RUIN DER STADT L.

Lisbon

Sited on what is perhaps the most magnificent natural harbour in Europe, where the River Tagus (Tejo) forms an inland lake just before it enters the Atlantic Ocean, Lisbon was once, albeit briefly, the maritime centre of the world. Its ancient name *Olisippo* may have suggested the picturesque legend that the city was founded by *Ulyssippo* – Ulysses. There is certainly evidence of settlement from around 1000 BC, probably as a trading post used by Phoenicians and then by Carthaginians, before the Romans chose the hill of Sao Jorge, just a few hundred yards from the water's edge, as the nucleus of their town. The Romans were ousted by the Visigoths, and the Visigoths by the Moors, who occupied the whole region in the year 714, and who remained for four centuries. In 1147 the Moorish citadel finally succumbed to a siege by a crusader army, comprising Portuguese, English, Flemish and Norman soldiers. The district of Alfama below the citadel still preserves its haphazard medieval winding streets.

It has been suggested that, when the reconquest of the Iberian peninsula was complete, the military, crusading spirit of the Portuguese and Spanish had to be re-directed outwards, and that this was the motive-force of the age of exploration. By the 1420s, the Portuguese had occupied the Azores and Madeira and they had begun a process of exploration of the coast of western Africa. Their motive was initially to find the source of the gold which was brought to Europe by Saharan Arab traders, but this aim was later replaced by the dream of circumnavigating Africa and finding a sea route to the Indies. This goal was finally achieved in 1498 when Vasco da Gama became the first European to reach India. Soon precious spices from south Asia were flowing into Lisbon and were being traded at a fraction of the prices demanded by the Venetians, who controlled the traditional route from Constantinople. Lisbon became briefly the focus of a new world, a trading empire which stretched 10,000 miles (16,000 km) from Portugal to the Spice Islands. Contemporaries described the city at this time as echoing to the sound of ship-building, awash with luxury imports from Asia, and thronged with ardent seafarers. This golden age lasted only a few decades: so many Portuguese joined the quest for gold and glory overseas that the domestic economy almost collapsed. Persecution of the Jews by the Inquisition forced many to emigrate, taking their money

Lisbon, 1756, by Matthaeus Seutter. A dual view showing first the old city with its palaces and churches climbing up the hillsides above the Tagus, and then the moment of the great earthquake with buildings collapsing into flaming ruins.
The British Library, Maps 11.e.2 (60)

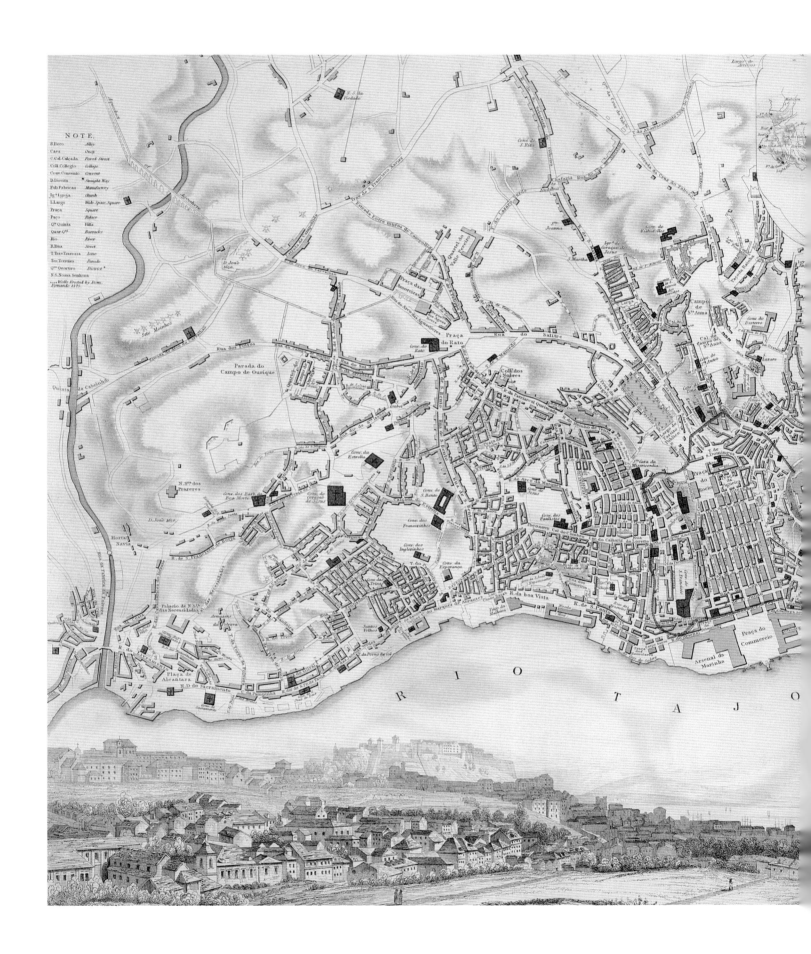

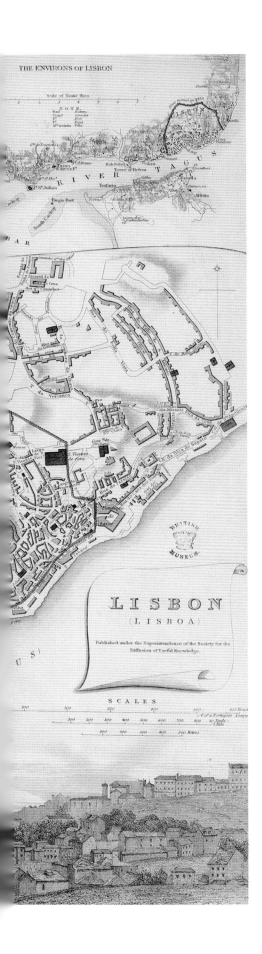

and their expertise with them. In 1578 King Sebastian and many of his nobles were killed in a disastrous invasion of Morocco, and Philip II of Spain annexed the throne of Portugal. Much of the overseas empire was lost to the Netherlands, and the most glorious era in Lisbon's history was over.

Lisbon remained nevertheless an important European capital, so that the events of All Saints' Day 1755 shocked the Christian world. This is how they were described by an eyewitness:

'On 1st November we had here very fine weather. In the morning at 9 o'clock I strolled on the Wall, then I went to the house of Mr. Cremer, and when I had been there for only a few minutes, the earth began to tremble. I did not know what was happening to me. All the people in the building fled, full of fear, into the garden. Hardly had we reached it when the roofs of the houses crashed down upon us. The earth moved so violently that we could not stand. The tremor lasted for almost twelve minutes, during which time the air grew so dark from the mortar and dust of the many crumbling buildings that I could not see the Sun. From all sides screaming and wailing could be heard. When the commotion had died down a little, I hurried full of fear and alarm back on board my ship anchored in the Tejo. The streets through which I had to pass looked pitiful; churches and monasteries had collapsed; people were lying in heaps, half dead and half alive. How happy I was to return alive to my ship. The whole city, as far as I could see, was almost destroyed.'

Estimates of those killed in this great earthquake are as high as 40,000, or one third of the city's population. Theologians and philosophers were at a loss to explain why such a slaughter should have been inflicted on a Christian city, many of whose people were in church at prayer when the earthquake struck. The earthquake was the great turning point in Lisbon's architectural history: under the direction of the Marques de Pombal, the city centre was rebuilt on a strictly geometrical plan. The line of the greatest destruction had passed through the area known as the Baixa, just west of the hill of Sao Jorge, and this devastated area was now completely levelled. From the new Praca do Comercio northwards to the Praca Rossio, a grid of streets was laid out with three-storeyed buildings, having tiled roofs, and eave-corners turned upwards; the height of the buildings was strictly uniform, as was the width of the streets. The style was a simplified baroque, to allow for rapid construction, and became known as *estilo Pombalino*. In the nineteenth century, the city extended for the first time northwards, the most striking addition being the opening of the wide boulevard, the Avenida da Liberdade. But the Baixa has been strictly preserved in all its regimented magnificence, while the Praca do Comercio remains the symbolic spot where Lisbon opens out to embrace the sea.

Lisbon, 1844. The grid layout of the central Baixa, rebuilt by the Marques of Pombal after the earthquake, can clearly be seen in this mid-nineteenth-century plan.
The British Library, Maps 38.e.8

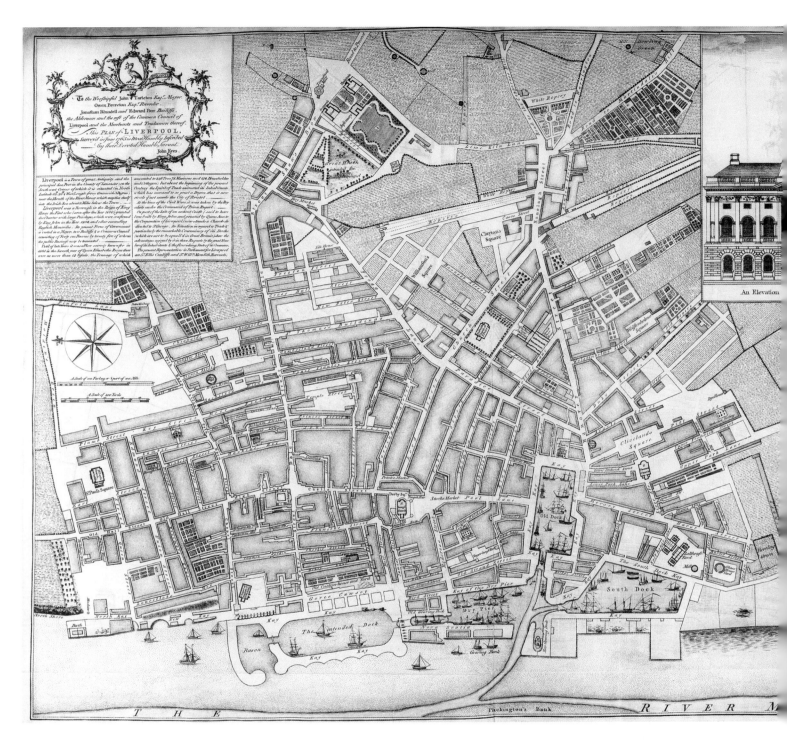

Above: Liverpool, 1765, by John Eyes. Looking east across the Mersey, this plan shows the extent to which the old docks reached into the very heart of the city. In contrast to the totally haphazard pattern of the old streets, an orderly new residential district is appearing on the right, with fashionable names such as Upper Park Lane and Upper Frederick Street. The lengthy text is full of local patriotism, proclaiming that 'Until of late years, Liverpool was a place scarcely known, but about the beginning of the present century the Spirit of Trade animated its inhabitants, so that it rivals, if not exceeds, the city of Bristol.' The inset picture of the Exchange stands like a secular temple to the gods of commerce.
The British Library, Maps K. Top. XVIII.71

Right: Liverpool's waterfront in 1797:
a superb artistic evocation of the
pre-industrial port.
The British Library, Maps K. Top. XVIII 76c

Liverpool

Unknown to the compilers of the Domesday Book in 1086, two centuries later Liverpool was a small fishing port and market town, beginning to trade with Ireland. The 'Pool' on which it was centred was a small tidal creek, since filled in and built over, emerging into the River Mersey in the area of the Salthouse Dock. The middle of the town consisted of Old Hall Street and Castle Street, and contained two market crosses where sellers of corn, meat and fish would gather. Overlooking the Pool mouth, a small castle was built around the year 1250, located where the court building now stands. South of the Pool were common lands, where cattle were grazed and crops grown. The map of Liverpool remained essentially unchanged for four centuries. It did not become a manufacturing centre of any kind, and even its sea-trade must have been very limited, for the first dock was not constructed until 1710, at which time the population was barely 6,000. The principal water supply was a well near the present St George's Hall, from which water was distributed in carts and sold at a halfpenny a bucket.

It was in the eighteenth century that Liverpool's fortunes changed dramatically on the back of the 'triangular trade' with Africa and the Americas. Cargoes of small manufactured goods were carried to the Guinea coast to be exchanged there for slaves, who were shipped to the West Indies and the American colonies. The return voyages brought sugar, rum, molasses, cotton and tobacco. These lucrative three-way voyages lasted about a year in all, and brought great prosperity to the merchant elite, incidentally damaging Bristol's trade, for the Mersey was far better adapted for larger ships than the Avon. Liverpool grew enormously, to a population of almost 100,000 in 1801, a rough, boisterous, seafaring town, its ethos conditioned by the hard, turbulent life of the sea, of smuggling, privateering and slaving, an ethos which it has never entirely lost. It is not surprising that the movement for the abolition of slaving met with fierce opposition and predictions that the town would be ruined by it. In 1807, the year of abolition, there were reportedly 185 Liverpool ships capable of carrying 144,000 slaves.

But there was no ruin: instead trade was diversified, with the tonnage of imported cotton for the Lancashire mills rising into the millions. New docks were constructed, and ship-building, metal-founding and all the allied trades prospered, so that Liverpool became a city of the industrial revolution, and the principal port of entry and departure for the Americas. The population was swollen by waves of immigration from Ireland, notably during the Irish famine years of 1845–48. The urban fabric and the social conditions in these years were appalling, with an astounding fifty-one per cent of children dying before the age of ten. The town spread rapidly but without any apparent system, as the street map still shows. Residential quarters were pushed more and more to the outskirts, and upper-class dormitories were built across the Mersey, on the Wirral peninsula, reached from 1820 by regular ferry services, and by the Mersey rail tunnel in the 1880s. A city at once both rich and squalid, humorous and tragic, proud and guilt-ridden, Liverpool was the creation of its seafaring trade. Its maritime and industrial past are long over, and although its waterfront has acquired a certain rugged beauty, its ambivalent heritage remains.

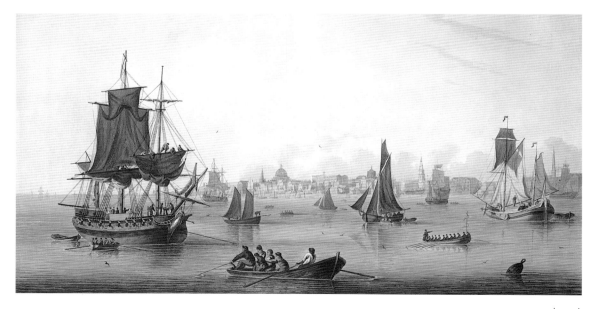

London, the influential and much-imitated engraving by
Visscher, published in 1616. Sea-going ships are confined
downstream of London Bridge, for centuries the only
crossing of the Thames, lined with houses and shops and
the heads of criminals on spikes above the gates. West of
the bridge are the Globe Theatre, newly rebuilt after the
fire of 1613 (caused by a spark from a cannon during a
performance) and the Bear Garden. Old St Paul's
dominates the north bank.

The British Library, Maps 162.o.1

London

At the first point where the River Thames could be bridged upstream from the marshes, the small city of London was built by the Romans in the first century AD. The river offered access to the interior of England and to the coasts of Europe; it was the natural capital of Roman Britain. Some three miles of boundary walls enclosed an area of 325 acres virtually identical to the area covered by the City today. We know almost nothing about the street plan of the Roman town, but it contained a military fortress, at least one temple, a forum and a basilica - a town hall, law court and place of public meetings. Almost certainly depopulated after the withdrawal of the legions in the fourth century, it was never abandoned, despite being taken, presumably by force, by the Saxons, and then suffering by fire and sword at the hands of the Viking raiders.

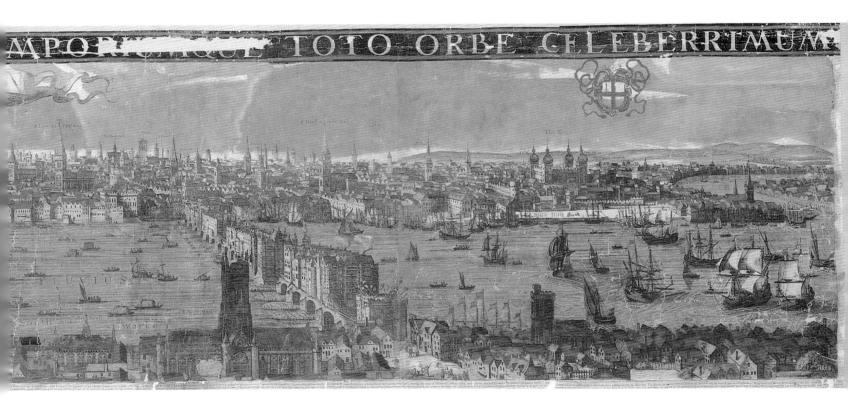

London's size and importance can be gauged from the tribute of £10,500 which the city paid to the Norse king Cnut around the year 1030, representing one eighth of the sum paid by the whole of England. A crucial event of the tenth century was the decision by the Saxon kings to build an abbey and a royal palace at Westminster, just two miles upstream from London. The Saxon capital remained at Winchester, but it would soon be drawn to Westminster. William the Conqueror sensed that London must be his seat of power, and here he was crowned and built his English fortress - the Tower. The Roman walls and the seven gates of London were restored, and with its churches and monasteries, merchants and craftsmen, markets and fairs, London was soon being spoken of by the medieval chroniclers as one of Europe's great cities.

Westminster provided a second focus for urban growth, and it was inevitable that the area between the two centres should be settled and built on. This process was well advanced by 1500, with many fine houses, some with riverside gardens, along the Strand. The Temple, home to the legal profession, already stood just outside the City's western walls, on property taken from the suppressed order of the Knights Templar. The amorphous spread of London beyond the City's ancient walls, the search for air and space, had begun. When Henry VIII built St James's Palace in 1531, a third focus was created which would in turn become linked with Westminster. Across the river, Southwark

London after the fire: one of many maps and views published in 1666 showing the void at the heart of the city, with just individual buildings islanded in the devastation.

The British Library, Maps Crace I.50

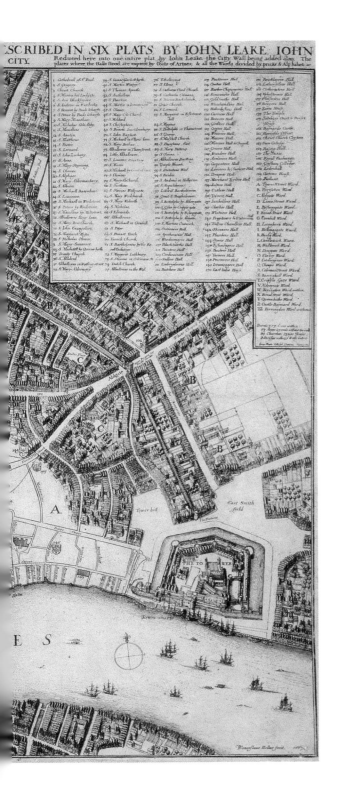

was the first 'suburb', through which the roads led south and east to Canterbury and Dover. The south bank was outside the City's jurisdiction, and it soon acquired a reputation for sport, crime, drinking and theatre. Tudor London also saw the first signs of ribbon development immediately outside the City gates: built-up streets now extended beyond Bishopsgate and Aldersgate into the open fields. During the Civil War, a line of ramparts had to be hastily thrown up by the Parliamentarians to defend this new London, which by then consisted also of Westminster, St James's, the Strand-Holborn area, Southwark and Lambeth.

The most dramatic events in London's history were the Plague of 1665 which killed almost 100,000 inhabitants, and the Great Fire of the following year, in which around eighty-five per cent of the City was destroyed, and some areas beyond it. In fact the Fire's long-term effects were few: the street-plan was virtually unaltered in the rebuilding period, despite the grand and imaginative plans put forward by Wren and others. The great memento of the Fire is St Paul's Cathedral, the masterpiece of English baroque which replaced the medieval structure that had been destroyed. In the new political and social atmosphere of the Restoration, the Strand emerged as an area of neutral ground between Court and City, the land of coffee houses, shops, theatres, taverns and printing-presses which it has always remained. Further west, many aristocratic landowners began to develop their estates, creating streets and squares of fine houses – Soho, Golden Square, Hanover, Cavendish, Grosvenor and Berkeley; the concept of the West End was born, although as yet there was no real East End to counterbalance it. Years of political peace and wealth derived from trade spurred the inexorable growth of London in the later eighteenth century. Between 1760 and 1766 London's seven gates and most of its ancient walls were demolished and the roads widened – a visible symbol of the city's breaking its historical mould. In Westminster and St James's, the Royal Parks remained thankfully inviolate, however. After a thousand years the river was at last awarded new bridges, beginning with Westminster in 1750, and with a dozen more being added until 1894 when Tower Bridge was opened. This period from the Restoration to 1760 is vivid to us from London's neoclassical architecture, and through the personalities associated with it – Wren, Pepys, Newton, Purcell, Defoe, Fielding, Handel, Hogarth and Johnson. The elegance of London's new streets and squares was entirely the creation of piecemeal private design and private investment: there was never any central control or overall plan for the capital.

The Industrial Revolution hit London with the new docks, the first built in 1802, and the associated mushrooming of the East End to house the population of dock-workers, sailors, and all the associated trades and industries. This process was hastened by the coming of the railways from the 1830s onwards, which dispersed population from the centre and brought all London's scattered districts into some kind of unity. The railways and then the underground created the suburbs, although the suburb of one generation could look distinctly central to the next, as they and the older, outlying villages – Greenwich, Chelsea, Putney, Hampstead – were gobbled up, and the metropolis grew haphazardly, still with no planning, reaching a population of five million by 1891. Mercifully, its growth was halted at certain key places by the work of the

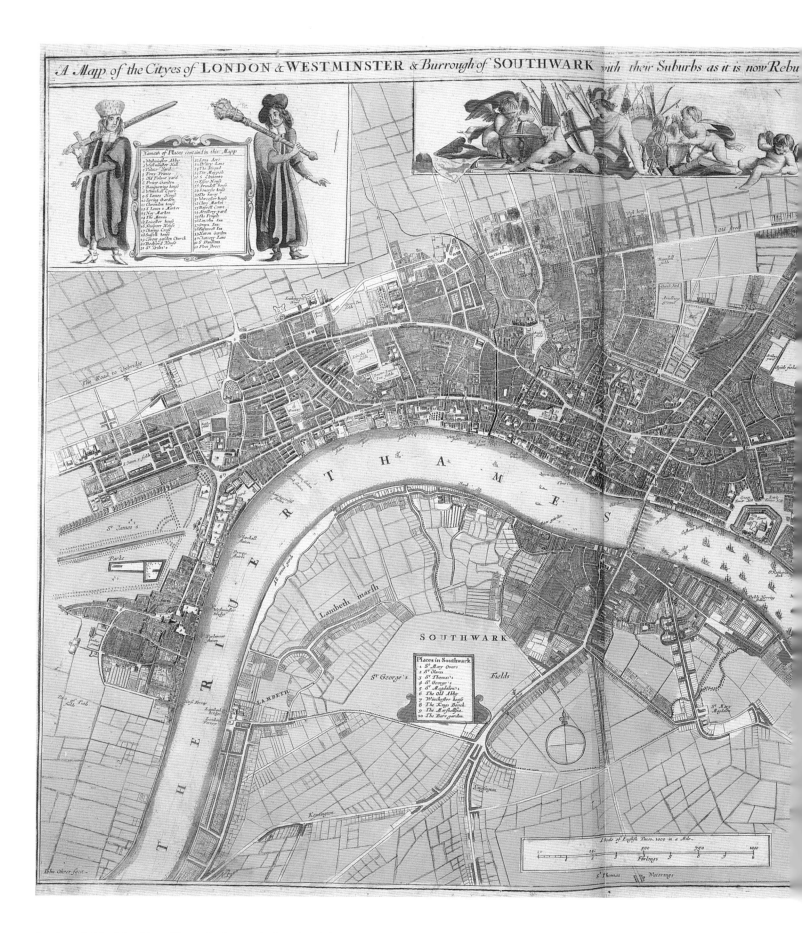

London, 1676, by John Oliver, 'Now rebuilt since the late dreadful Fire'. One of the incidental effects of London's Great Fire was that it gave birth overnight in England to the scaled town-plan, which was essential to any rebuilding plans, and which now emerged to replace the traditional, picturesque bird's-eye view.

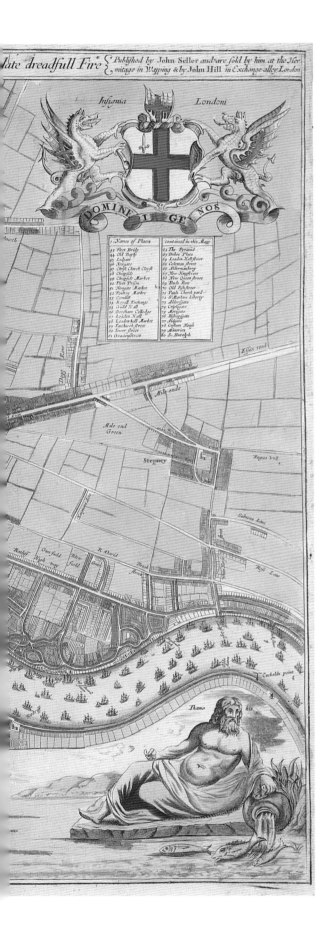

Commons Preservation Society, who saved areas such as Hampstead Heath, Wimbledon Common and Epping Forest from development. To its citizens London was then the 'Heart of the Empire', the centre of government, finance, international trade, the arts and social life; it was without equal among the world's cities. The public and private architecture of the elite was magnificent; the slums where the poor lived were soul-destroying. This was the legacy with which the new democratic councils of the twentieth century began to grapple, and to which was added the effects of the revolution in transport which challenged historic cities throughout the world.

This is the external history of London's growth, the official story. But it never explains the living city, how so many Londons have come into being, every few streets emerging into its own village with its own personality, memories and ghosts: Covent Garden, Blackheath, Whitechapel, Pimlico, Canonbury, Leytonstone, North Woolwich, Herne Hill, Barnes – the list goes on and on, some famous and fashionable, others not mentioned in any guidebook. The past is everywhere in London, but the physical and social environment is changing so rapidly that it is essentially the past of the mind, with one's own childhood almost on the same now-mythical level as images such as 'Dickens's London'. Many people have hated London and have evoked its desolation in Blakean terms, but even Wordsworth, the supreme poet of nature, could not resist this sense of the city's personal force, that some power derived from its history was latent in the streets and bridges, in the flowing river, and in the very air. How this vitality can have come into being without planning, without democracy, without intention, is the great enigma of urban history.

Heart of Empire by Niels Lund, 1901; a painting whose title reflected London's importance on the world stage at the turn of the twentieth century.
Guildhall Library, Corporation of London

Luxembourg, besieged by the French army in 1684. A beautifully detailed etching by Romein de Hooghe, the Amsterdam artist. The fires and destruction are clearly visible, and the artist has perhaps shown his own feelings about the event in the symbolic cartouche, where the warlike figure of France is seen banishing justice, and whose text stresses the violence of the battle.

The British Library, Maps C.9.e.4 (63)

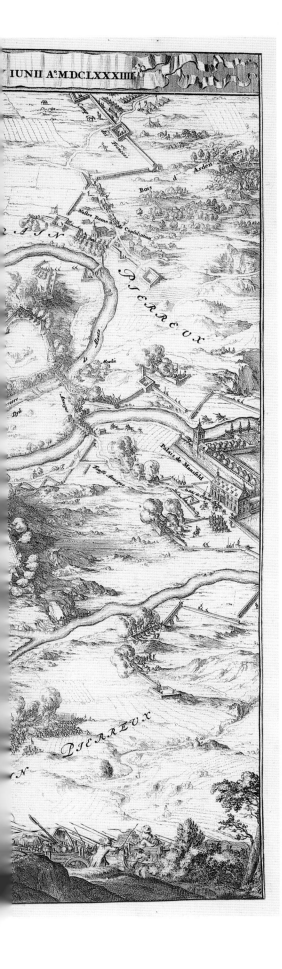

Luxembourg

Squeezed uneasily between France and Germany, the identity and survival of Luxembourg as a state would have been unthinkable without its fortress, set on a high sandstone ridge above a loop in the Alzette River. This ridge had been fortified by Romans and Franks, and in 963 AD it became the seat of an independent duchy. The castle itself was demolished in 1867, but Luxembourg remains a fortress city built on two levels, with the lower riverside suburbs dominated by the central historic hilltop.

This view of Luxembourg from the north-east shows the city under siege by the French army in 1684. At that date the Grand Duchy had been for two centuries part of the Hapsburg dominions, and Louis XIV's attack on it was part of his strategic battle to limit Hapsburg power and steadily expand his own national boundaries. The castle had been fortified with the typically massive seventeenth-century bastions seen here, and the inhabitants resisted the siege from 19 April until the final capitulation on 4 June, after a ferocious bombardment by the French artillery. The city is shown here entirely surrounded by French cannon, although the main attack clearly comes from the open ground to the south, where the large French encampment can be seen. Although the city suffered great devastation, some of these bastions are still standing. No long-lasting results came from this destructive siege: France returned Luxembourg to the Hapsburgs in 1697, in one of the periodic truces in King Louis's wars with his neighbours.

Independence was restored in the post-Napoleonic settlement of 1815, and it suited the European powers to foster a small, neutral, independent state between France and Germany. This neutrality was violated in both World Wars, but this historic position was again confirmed by its role as seat of some of the founding institutions of the European Community.

Luxembourg, a distant view of the city in the year of the French siege. Close inspection of this apparently placid view reveals a city half in ruins. It shows, too, the steep cliffs over the River Alzette on which the city is built.
Giraudon/www.bridgeman.co.uk

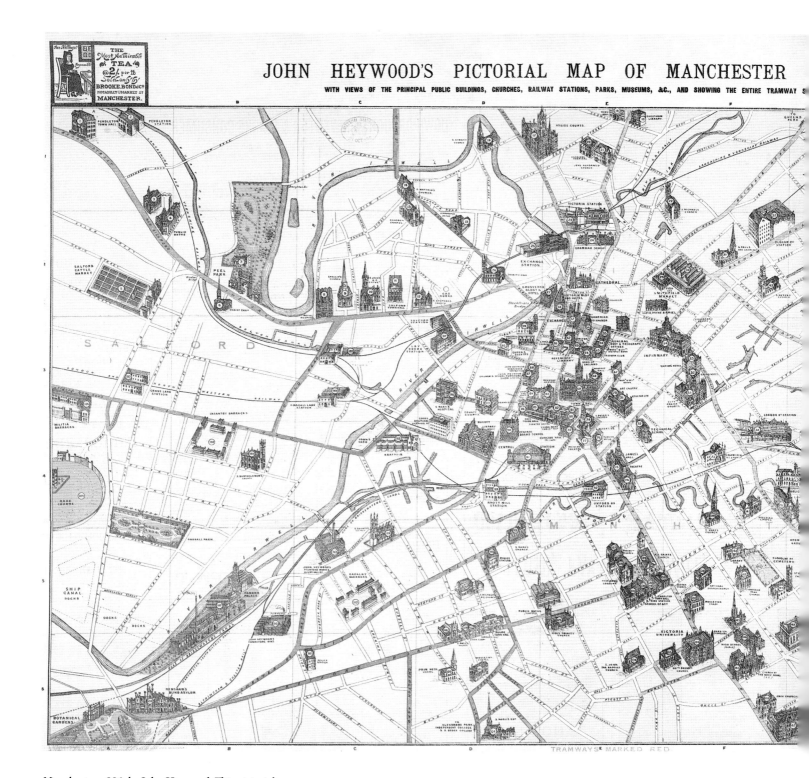

JOHN HEYWOOD'S PICTORIAL MAP OF MANCHESTER

WITH VIEWS OF THE PRINCIPAL PUBLIC BUILDINGS, CHURCHES, RAILWAY STATIONS, PARKS, MUSEUMS, &C., AND SHOWING THE ENTIRE TRAMWAY S

Manchester, 1886, by John Heywood. This pictorial
map conveys, no doubt unintentionally, something
fundamental in the character of Victorian Manchester.
The towering architecture of the public buildings rises
proudly out of an apparently empty plain: the houses
where the common people lived, and the factories
where they worked, have been mysteriously wiped
off the map, for fear of spoiling the view.
The British Library, Maps 3215 (16)

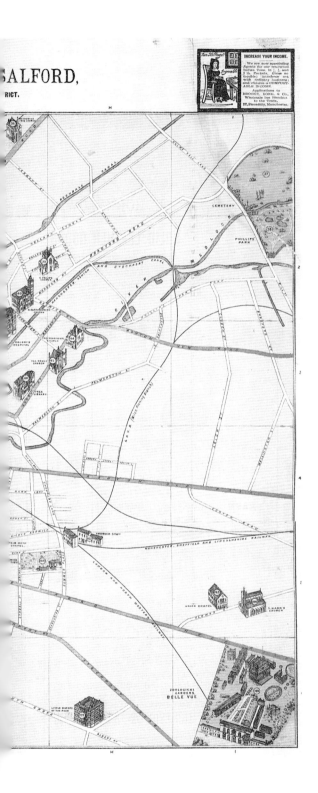

Manchester

Manchester was the archetypal Victorian city, the scene of explosive population growth, technical innovation, monumental civic architecture and appalling squalor. It was the prototype of an entirely new phase of urban history – the first of the industrial cities which were to become the norm throughout nineteenth-century Europe, then America and then the world. As early as 1724 it had been described as 'the busiest village in England', having then a population of some 10,000, most of whom were engaged in the weaving of woollen and cotton fabrics, for which the place had been famous since the Middle Ages. It was in Manchester that the great revolution in the commercial life of England – the development of the factory system – began in the 1780s, when handwork in the individual home was replaced by mass-production in huge mills, where the workers served the new steam-driven weaving machines. The Bridgwater Canal had already brought coal from Lancashire to the town to power the mills, and had given the town an outlet to the sea. The population rose to 75,000 by 1801, with almost as many more in a ring of cotton towns which surrounded the centre – Bury, Bolton, Rochdale and Oldham; as these towns were gradually engulfed, so Manchester would become recognised as the first 'conurbation'.

Technical innovation and wealth creation were the essence of the town's existence, but these historic changes exacted a high human cost. The town had no system of democratic government; it was managed like a village, through a manorial court, and sent no representative to Parliament. It provided classic ammunition for the advocates of political reform between the years 1790 and 1830. The Peterloo Massacre, when a dozen citizens were killed and hundreds injured by armed troops, occurred during a political rally demanding Parliamentary reform. Only in 1838 did Manchester acquire a charter and an elected governing council. By 1851 the population topped 300,000, and the living conditions of many were vile. In the 1840s the German social philosopher, Friedrich Engels, lived in Manchester, and his classic work *The Condition of the Working Class in England*, 1845, was based on his experiences there. The novels of Elizabeth Gaskell were also set in Manchester in the hungry forties, and were the first fictional works to unveil the grim reality of industrial society; they were deeply disliked by the mill-owners and by the Conservative press, but inspired Dickens. The environment in Manchester at this time was poisonous: industrial smoke intensified the region's natural cloud-cover, housing was overcrowded and insanitary, and the River Irwell was effectively the open sewer of the entire town.

In spite of this dark setting Manchester, or 'Cottonopolis' as it was known, was a dynamic city, which displayed its wealth proudly in monumental architecture. Banks, warehouses, offices and public buildings were clothed in superb neoclassical or Gothic forms. Public squares were adorned with statues of nineteenth-century heroes: Wellington, Peel, Cobden, James Watt and the Prince Consort. Philosophic and political clubs abounded, and a vigorous press grew up. Cultural life flourished through libraries, museums, orchestras, and the first of England's great civic universities. All this was a conscious statement that culture could thrive side by side with commerce. Engineering came to rival spinning as the leading industry, while the opening of the great ship canal in 1894 brought seagoing ships into the heart of the town. The decline of Manchester's industrial base in the twentieth century led to a protracted crisis, from which a regenerated city is just emerging.

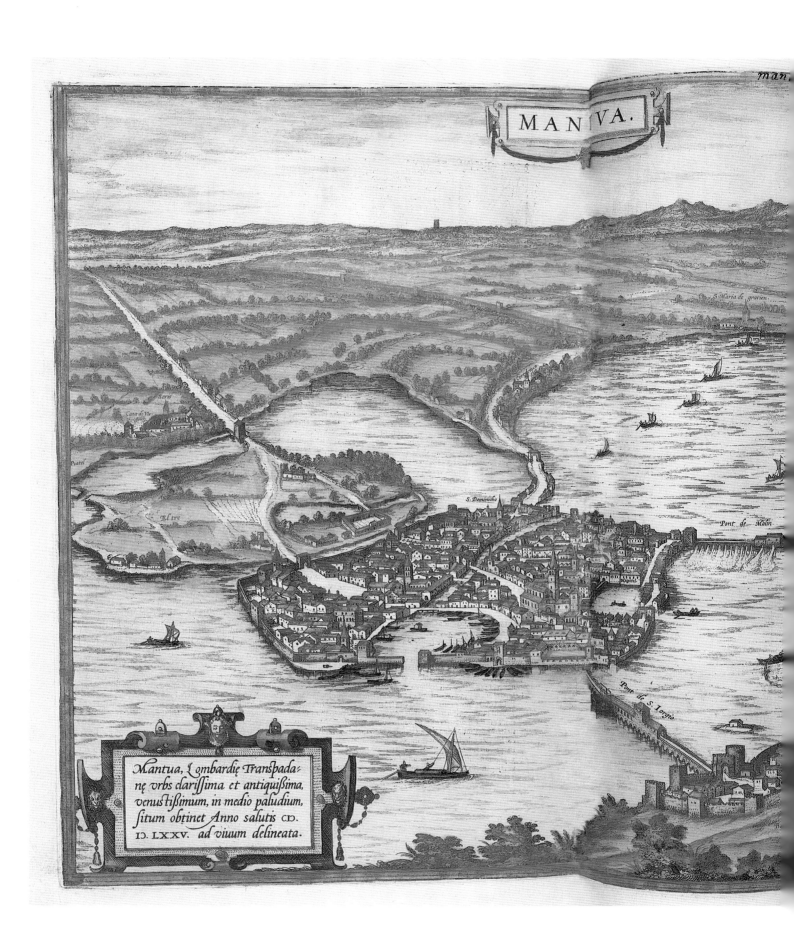

MANTVA.

S. Maria de gratien.

Pont de Molin.

S. Dominicho.

Pont de S. Iorio.

El tre.

Mantua, Lombardię Transpada-
nę vrbs clarissima et antiquissima,
venustissimum, in medio paludium,
situm obtinet Anno salutis CIꓛ.
Iꓛ. LXXV. ad viuum delineata.

Mantua

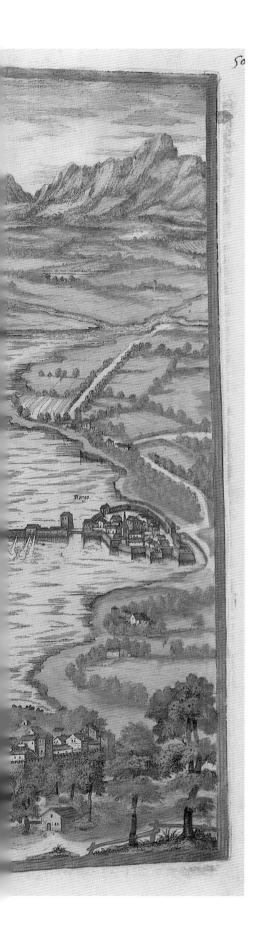

Perhaps the most distinctive of the smaller north Italian cities, Mantua occupies a peninsula site between lakes formed by the River Mincio, and a sizeable canal bisects the town centre. A Roman town, famed as the birthplace of Virgil – always known as 'the Mantuan' – it was virtually destroyed by Alaric and Visigoths in 401 AD, and several centuries of obscurity followed. Mantua's golden age came between 1400 and 1600, when it was home to the Gonzaga family, whose humanist court became one of the most brilliant of the Renaissance, earning it the title 'the New Rome'. The Gonzagas employed first Mantegna then Giulio Romano to decorate their palaces, while other figures patronised by the court included the composer Monteverdi, whose early operas were all performed here, Castiglioni the political writer, and the epic poet Ariosto. Between Virgil's time and the Renaissance the city boasted another celebrated son, the troubadour poet Sordello, subject of Browning's famously obscure poem.

The Gonzagas were feudal warriors who diversified into commerce and banking, although many of them retained their soldierly traditions. It was Ludovico Gonzaga who brought Andrea Mantegna to the city in 1459 to decorate the Castella de San Giorgio. Mantegna's series of group portraits in 'Wedding-Chamber' and the 'Painted Chamber' set out to create an illusion of total space around the walls, which must have functioned like a mirror to the members of the family, who would have seen their own images gazing back at them. Very different from this aristocratic art was the work of Giulio Romano, whose art was a debased version of that of his teacher Raphael, and his opulent frescoes all take pagan themes, enlivened with many obscene details. Romano's fame in his own day was immense – he is the only artist mentioned by name in a Shakespeare play (*The Winter's Tale*). The splendour and intrigue of Federigo's court during these years form the setting for Verdi's opera *Rigoletto*. Within a few years the city's population reached 40,000, swelled by a large Jewish community which found refuge from the Inquisition in the indulgent regime in Mantua. The streets spread far beyond the *citta vecchia*, which was centred around the cathedral and the San Giorgio bridge, expanding always south and west, for the lands across the broad river were never settled.

By the seventeenth century a decline set in. Duke Vincenzo II and his successors squandered the family wealth and corrupted the government of the city. The population fell, the fields around were neglected, and the golden age was gone. In 1708 the last duke died childless, and in the same year Mantua was occupied by the Austrians during the War of the Spanish Succession. It remained under Austrian rule until 1797, when it fell to Napoleon after an eight-month siege. In 1814 it was returned to Austria, but Italian nationalism found a ready home in Mantua, and it joined the new Kingdom of Italy in 1866.

Seen from the lake, the city's skyline is almost as impressive as that of Venice, and today the centre of Mantua is a compact, aristocratic memorial to a long-vanished era, when the energies of a single family could shape the destiny of a whole city for three centuries. Mantua's history mirrors the fall from grace suffered by so many of the Italian city-states: from Renaissance wealth and splendour to years of submission to foreign powers. Only when they merged their identity into that of Italy did they emerge from their slumber.

Mantua, 1572, by Braun and Hogenberg. This view from the north-east across the San Giorgio Bridge emphasises the city's status as a virtual island, almost Venetian in appearance. Strangely, perhaps, the artist has rather understated the ducal palaces so that they are not easy to see: the Palazzo del Te stands across the southern moat – the moat which has long been in-filled and become the Viale del Risorgimento. The basin between the two northern bridges is now a park named after Virgil, Mantua's most famous son.
Birmingham Libraries

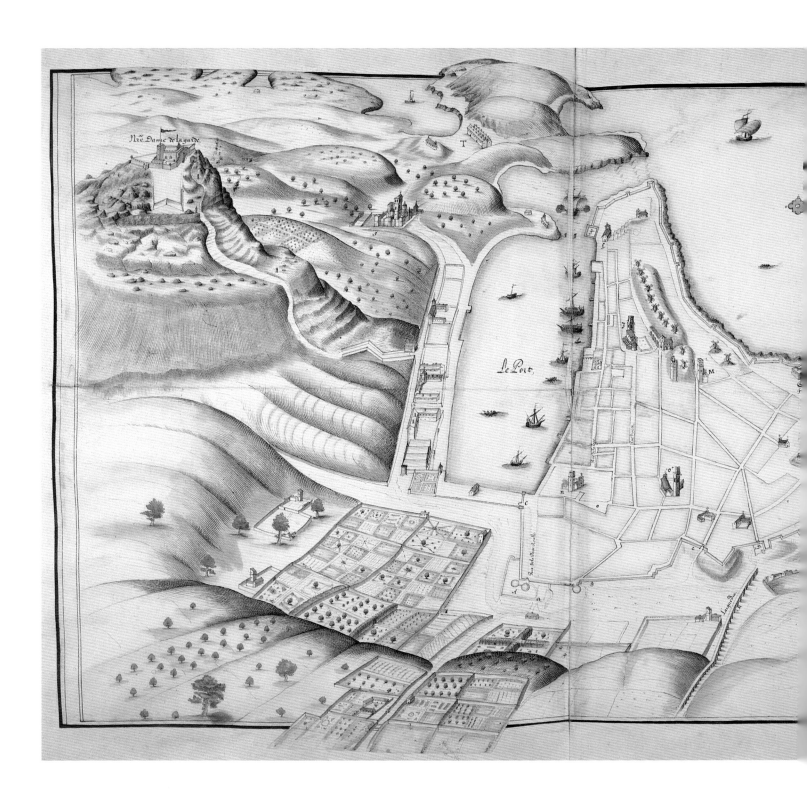

Nre Dame delagarde

Le Port.

Marseille

Marseille, in size the second city of France, is by far the oldest: a Greek colony established around 600 BC, it is several centuries older than Paris. The deep harbour in the otherwise sandy or marshy coastline was always its great asset, and maritime trade flourished to the extent that 'Massilia' became the parent city of further colonies at Monaco, Nice and Antibes. It was from Massilia that the Greek seafarer Pytheas set out in the fourth century BC on the first voyage of discovery to northern Europe, circumnavigating Britain and perhaps crossing the North Sea to Scandinavia. The Romans took over the town and built roads linking it to the Italy–Spain route at Arles and Aix.

In the post-classical era, the city crumbled into near extinction, pillaged by Goths, Muslims and raiding pirates. Revival came in the twelfth century, largely because Marseille functioned as the port of disembarkation for the crusaders. The city prospered, and in the thirteenth century it declared itself a republic within the kingdom of Provence. The harbour was protected by the twin towers where the Fort St Jean and the Fort St Nicolas now stand, and a great chain was drawn up each night across the entrance. The port buildings and dwellings were concentrated to the north of the harbour, in the Panier quarter. Outside the walls industries such as tanning, weaving and soap-making sprang up, and the spread of the city was haphazard. During the Fronde in the 1650s, when the provinces of France were resisting royal absolutism, Marseille rose against Louis XIV, who came in person in 1660 to breach the city walls and subdue the rebellion. It was only after this date that the Marseille which we now see began to take shape, with the building of a new quarter to the west. The great boulevard of La Canebière was laid out, with diagonal streets converging in the classical pattern, and some fine new buildings on the Cours Belsunce. In 1720, however, the city was decimated by an outbreak of plague, brought by ships from the eastern Mediterranean. Two of the islands out in the bay, Ratonneau and Pomegues, had long been used as quarantine stations, but on this occasion they failed to protect the city and reputedly half of the population of 100,000 perished.

Marseille joined enthusiastically in the Revolution, and its band of volunteers marched on Paris singing a battle song which thrilled Paris and at once became known as the Marseillaise. During the Terror however, the city rebelled against the Convention, and was again subdued by armed force. Marseille was then for some years fiercely anti-Bonapartist, and supported the restored monarchy. This period was the setting of the book which was once indelibly identified with Marseille, *The Count of Monte Cristo*, whose opening scenes in the port and in the infamous island-prison of the Château d'If thrilled millions of readers two or three generations ago. The Château was built on the third island in the bay by Francis I, but by the eighteenth century it had become, like the Bastille, a private prison where the innocent could be immured without trial under royal warrant. Dantes's escape from this fortress, sewn up in the shroud of his dead friend, is one of the immortal episodes of fiction.

During the nineteenth century, Marseille expanded hugely as it became the port of the French empire in Africa, and, with the opening of the Suez canal in 1869, in the Far East. From a population of 100,000 in 1800, it rose to half a million in 1900, and today to one million. The new port of Joliette to the north of the old harbour was excavated in the 1860s. Marseille suffered huge damage in World War II, when the entire Panier district was dynamited by the Germans, partly in reprisal for Resistance activities. Marseille today reveals little of its 2,600 years of history. It is a cosmopolitan, sometimes violent, and functional city. Seen from the hill of Notre Dame de la Garde (from which, for centuries, approaching ships were signalled), it sprawls endlessly beyond the heartland of the old port. Marseille's conception of itself has always been functional, as a port, a place of work, not a place of art or fashion. Yet no city in France is more alive, and it has at least been spared the ravages of tourism.

Above: Marseille, a manuscript view c.1602–8, from the east towards the hilltop basilica of Notre Dame de la Garde. The old port is confined to the north side of the harbour, while open fields still surround the city to the east and south. At the entrance to the harbour, the two forts of St Jean and St Nicolas can just be seen, between which a great guard-chain was slung each night.
The British Library, Add MS 21117, ff.94v–5

Left: Marseille: the Chateau d'If, the island prison just outside the harbour, made famous by Dumas in *The Count of Monte Cristo*.
The British Library, Harley MS 4421, ff.13v–4

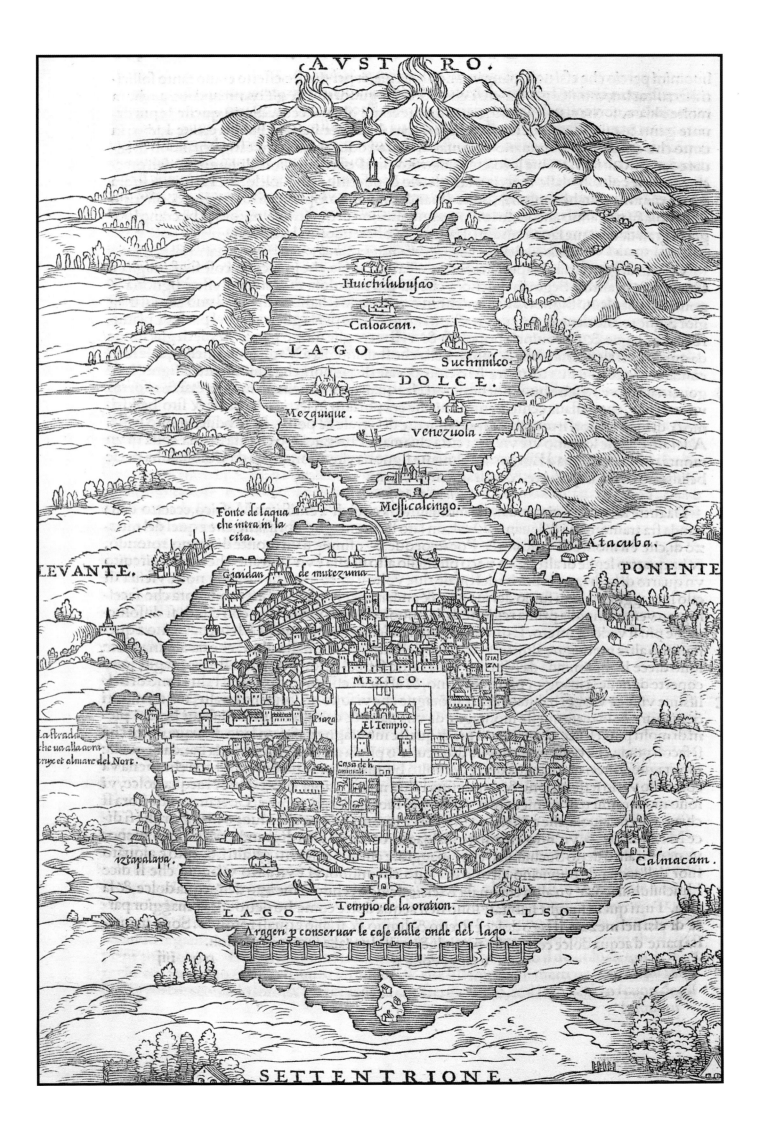

AVSTRO.

Huichilubufao

Caloacan.

LAGO

Suchimilco.

DOLCE.

Mezquique.

Venezuola.

Messicalcingo.

Fonte de laqua
che intra in la
cita.

Atacuba.

LEVANTE.

Giaidan   de mutezuma

PONENTE

MEXICO

Piaza

El Tempio.

La strada
che va alla avra
cruze et almare del Nort.

Casa de li
animali.

iztapalapa.

Calmacam.

Tempio de la oration.

LAGO

SALSO.

Arageri p conservar le case dalle onde del lago.

SETTENTRIONE.

Mexico City

'And when we saw all those cities and villages built in the water, and other great towns on dry land, with that straight and level causeway leading to Mexico, we were astounded. These great towns and buildings rising from the water, all made of stone, seemed like an enchanted vision ... It was all so splendid, this first glimpse of things never heard of, seen or dreamed-of before.'

So wrote Bernal Diaz, an eyewitness of the Spanish conquest of Mexico. Mexico City stands today on the site of the Aztec city of Tenochtitlan which Diaz was describing, breathtakingly conjured, like Venice, out of the midst of a lake. It had been established around the year 1325, when the migrating Aztecs had witnessed on the island what they took to be a sign from their gods to settle there: an eagle perched on a cactus devouring a serpent. In the 150 years which followed, the Aztecs had mastered the surrounding tribes and extended their city to the point where it was home to more than 100,000 people. Without the wheel and without metal tools, the Aztecs were great structural engineers: the island was linked to the shore by a number of causeways; transport within the city was via a system of canals; the level of the lake was regulated by a flood-barrier; palaces and temples were built.

Yet at the heart of this unique city lay a nightmare: the temple of the god Huitzilopochtli was a pyramid on whose summit human victims were butchered daily, often in their hundreds, in order to offer to the gods the blood which the Aztecs believed was needed to uphold the cosmic order. This practice fascinated and repelled the Spanish, who used it to justify their conquest of a barbaric nation. Only one authentic image of this island city reached Europe, drawn by command of Cortés himself and included with his famous letter to the Emperor Charles V describing the realm of the Aztecs. It was printed again and again, but it showed a city which had already disappeared: after a year's violent conflict, Tenochtitlan was taken and the Aztec rulers killed. Cortés then used the populace as slave labour to raze the city and erect a church where the infamous temple had stood. A colonial city was soon built, focused as the earlier one had been on the great plaza which was later christened the Zocala. Here, around the central market, the offices of the colonial government took shape, including the viceroy's palace, built on the site of that of the Aztec king. The use of ornate architectural motifs was facilitated by the abundance of soft volcanic stone called *tezontle*, and the Mexican baroque would reach its height in eighteenth-century structures such as the Metropolitan Cathedral. The new city was named Mexico, the *Mexicas* being one of several names for the Aztec people, and the country at once became New Spain. Many attempts were made to drain the lake, none of them wholly successful until the nineteenth century, so that both flooding and typhoid fever were endemic. The native population, however, was decimated by European diseases to which they had no immunity – influenza, measles and smallpox. To replace them, the Spaniards imported slaves from Africa.

Independence from Spain came in 1821, but the subsequent history of the city is an unending succession of conflicts and tyrannies, riots and revolutions. In 1847 a United States army occupied the city in the war which took Texas, New Mexico and California from the Mexicans. It was the Emperor Maximilian, a Hapsburg Archduke placed upon the throne by the European powers, who

Mexico City, 1556. All European representations of the Aztec capital were based on one sketch sent back to Spain with Cortés's original reports. This later version has been embellished by the background drawings of Mexican volcanoes. The almost Venetian appearance of the city, with its temples and palaces floating on the waters of the lake, amazed the conquistadors; but their admiration could not prevent their destruction of the city and, when this map appeared, the city shown here no longer existed.
The British Library, 566.K.1–3

restored the Chapultepec Castle to tower over the central city as the Acropolis does over Athens. Maximilian also laid out the grand avenue Paseo de Reforma, formerly Paseo Emperador, and this Haussmanisation of Mexico was extended during the long presidency of Porfirio Díaz from 1876 to 1911. During the Mexican revolution of 1910–17, the city became a battlefield, as waves of displaced peasants poured into its streets. In 1920 the population was barely half a million, but now Mexico City is probably the largest city in the world, with social and environmental problems on a massive scale. Mexico once devoured people, but now seems to spawn them, and the population is reputedly twenty million. There is violence and tragedy inherent in Mexico's past. No other capital city in the world is built, as Mexico was, on the ruins of a shattered civilisation, while few other cities embody so painfully the ills of modern urban living.

Mexico City: the Plaza Mayor and the baroque cathedral, standing almost precisely on the site of the Aztec temple of Quetzacoatl.
The British Library, 648.c.1

Моscow from the Blaeu Atlas of 1660. This view from the east emphasises the successive concentric rings of city walls which grew outwards from the Kremlin and from the Kitay-gorod beside it. The two inner walls were of white stone, the third at this date was of wood. The small Neglina River, now vanished, had been channelled to create a defensive moat. As in most cities before the nineteenth century, only one bridge existed across the main Moscow River. Blaeu's very detailed plan must have been based on an authoritative Russian original, perhaps that drawn for Boris Godunov around the year 1600.

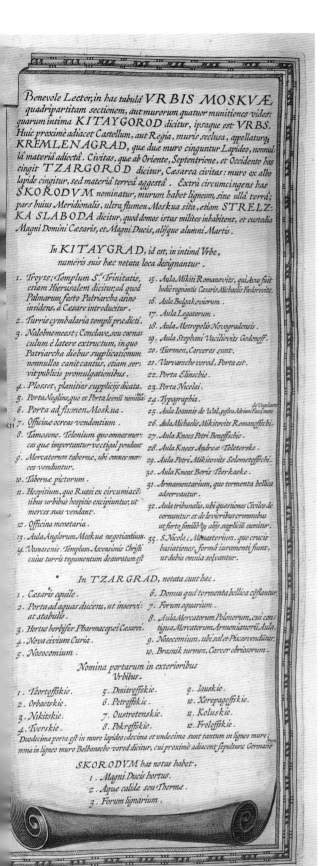

Moscow and St Petersburg (Leningrad)

The cities of Moscow and Leningrad-St Petersburg are the twin poles of Russian history: Moscow ancient, grim, religious, the heart of the country; Leningrad modern, magnificent and secular, conceived as a symbol of royalty, yet the birthplace of the Revolution. The official architecture of both cities is imposing, cold and authoritarian, and the names of both cities evoke images of winter, of cruelty and human suffering at the hands of invaders, but still more at the hands of their own tyrannical rulers.

The date of Moscow's origin is uncertain. Built on a slight rise where two small streams entered the Moskva River, archaeological evidence suggests that it was inhabited from the Neolithic period. It enters history in 1147, when the Kremlin functioned as the citadel of a regional prince; many Russian towns had 'kremlins' or fortresses, built at this time of wooden palisades. In the thirteenth century it was attacked and destroyed by invading Tartars, and was forced to accept Mongol suzerainty. The city was rebuilt and in 1326 the head of the Russian church transferred his seat from Vladimir to Moscow; the city's position as the centre of orthodox Christianity was strengthened after the fall of Constantinople in 1453, when it claimed the title of 'The Third Rome'. The Princes of Muscovy steadily extended their rule at the expense of other principalities, especially Novgorod. It was in the reign of Ivan III (r.1462–1505) that Moscow became the undisputed centre of a unified state. Ivan brought architects from Italy to remodel the Kremlin with new walls, and the churches of the Annunciation and Archangel were built in the ornate Byzantine style. To the east, the trading and market area known as Kitay-gorod flourished, where Red Square now stands. The Kremlin has always remained a magnificent and unique ensemble of buildings, which were public and religious, but standing cheek-to-cheek with the edifices of the authoritarian state. Outside the walls, the town was little more than a jumble of wooden huts, where, in the unflattering reports of foreign visitors, the inhabitants relieved their miserable existence with endless drinking and the dead lay in the frozen streets to be gnawed by dogs. The threat of Tartar invasion was still present until the late sixteenth century, and these outer villages were enclosed by successive ring-fortifications, whose location can still be clearly traced. Moats were dug around the Kremlin making it an island-fortress, while outside the city a ring of fortified monasteries was built, including the famous Novodevichy Convent, which played an important role in times of conflict – in the reign of Boris Godunov, of Peter the Great, and Napoleon's invasion.

The decision of Peter the Great to found the new capital of St Petersburg may have seemed a great blow to Moscow, but it did not halt the city's growth. New industries sprang up, the university was established in 1755, fine private houses were built by some of the same architects employed at St Petersburg, and yet another outer ring of fortifications was built, 25 miles (40 km) in length, still marked today by the many streets called *Val* - rampart. The population was almost 300,000 by 1800, still much larger than St Petersburg's, and in 1812 Napoleon knew that it was Moscow, Russia's heart, at which he must strike. The fire which the defiant Russians lit to frustrate him destroyed more than two thirds of the city, and it marks a watershed in the history of Moscow, for almost

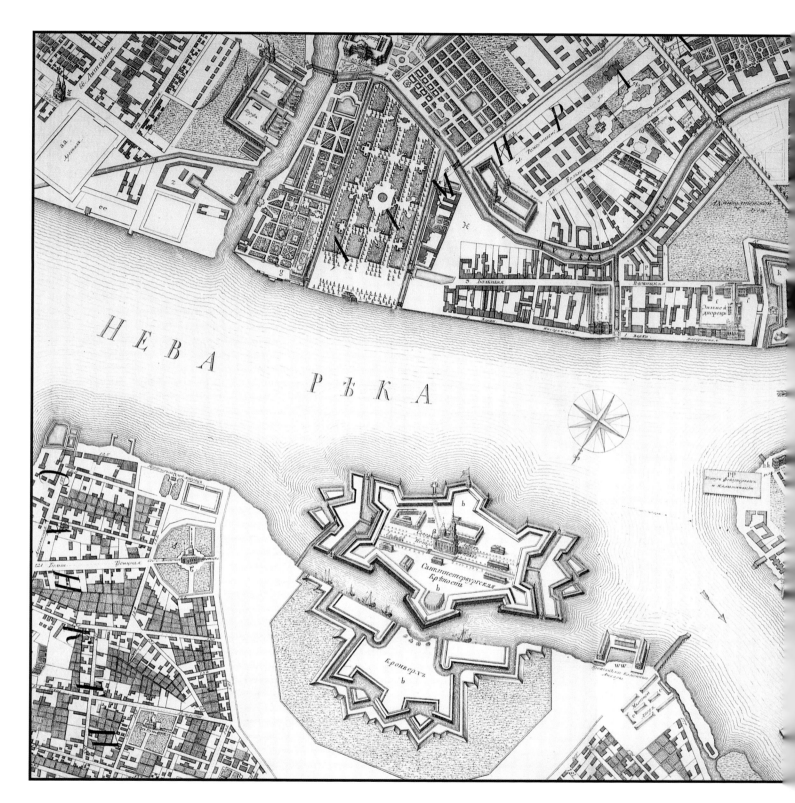

St Petersburg, 1753, with south at the top of the map.
The Peter-Paul fortress on its island faces the Admiralty
building and the Winter Palace. This is an interesting
transitional map, in which the streets are shown in plan,
but where the map-maker could not resist showing the
great buildings in elevation.

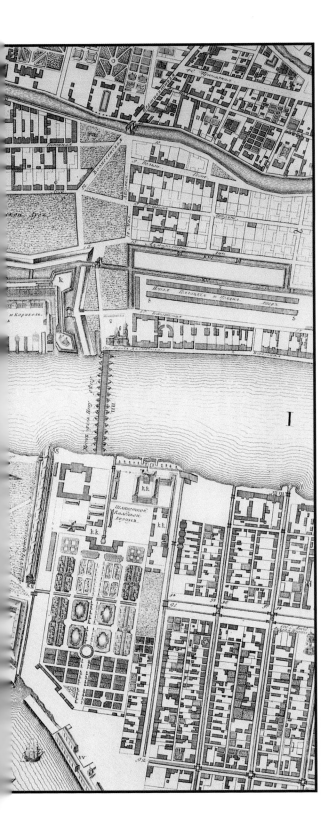

nothing outside the Kremlin pre-dates that fire. The rebuilding was rapid, and industrialisation coupled with the emancipation of the serfs in 1861 created a population explosion to almost two million by 1914. It was in Moscow that the parliament, the *Duma*, was established as a short-lived gesture to Russian democracy. St Petersburg was the cradle of the 1917 revolution, but almost immediately afterwards, the government returned to Moscow, which then became for seventy years the centre of world communism, its name symbolising an idea and the menacing shadow of a great power, just as Rome's had once done.

St Petersburg is unique in modern history as a city conceived and built by the decree of one man. The region known as Ingria, between Lake Ladoga and the Gulf of Finland, was taken by Tsar Peter the Great in 1703 from the Swedes who had occupied it for a century. Here, at the mouth of the Neva River, he determined to create a Russian gateway to the west. The site was marshy and unsuitable for building, but architects, craftsmen and labourers in their thousands were brought in to drive great foundation-piles into the earth. The Peter-Paul Fortress on Zayachy Island and the Admiralty shipyard on the opposite, southern shore of the Neva formed the core of the new city, which in 1712 was proclaimed the capital of Russia. Nobles were compelled to build houses on either bank or on Vasilevsky Island, while naval traffic was diverted from Archangel, Russia's long-established northern port, to the new harbour. Work was begun almost immediately on the series of waterways which would link the city to the lakes and the Volga basin to the east. In the early years everything had to be brought to the site, but by the 1730s metal-foundries and factories making weapons, textiles, paper, furnishings and china made the city self-supporting, although its growth continued long after Peter's death in 1725; he never lived to see the Nevsky Prospect for example, which was laid out after the 1720s. The most visible symbols of regal St Petersburg were the many palaces in and around the city: the Summer Palace and the Winter Palace in the centre, the Peterhof on the Gulf of Finland, Tsarskoye Selo, Pavlovsk and Gatchina to the south, all built in the Rococo style by Italian and Russian architects. Each new monarch apparently felt the need to create his own regal domain, and the post-revolution survival of all this royal architecture is most surprising.

Industrialisation was as rapid here in the years between 1850 and 1900 as it was in Moscow, and the seeds of revolution were sown in the appalling world of the factories, labouring beneath an autocratic regime. Insurrection in the city see-sawed with repression, culminating in January 1905 in the march on the Winter Palace, which ended in the massacre of hundreds of protestors. World War I saw the city's German-style name changed to Petrograd, but two years of military disaster brought the people to near-starvation, and in February 1917 food riots sparked the first revolution, and the formation of the provisional government

under Kerensky. On 3 April, however, Lenin arrived at the city's Finland Station from his long exile, and set about constructing a wholly new order. The Tsar and his family were imprisoned at Tsarskoye Selo before being sent to their deaths in Siberia. In October the Bolsheviks seized power by force, training the guns of the cruiser *Aurora* on the members of the Kerensky government, sitting in the nearby Winter Palace. With the government moved to Moscow, and soon renamed Leningrad, the city suffered terribly in the harsh years of the civil war followed by the Stalin era, but the worst was to come. From September 1941 to January 1944 Leningrad was besieged and blockaded by the German army; half the city was reduced to rubble and three quarters of a million people died, mainly of starvation. Shostakovich's brooding and desolate Seventh Symphony commemorates this experience. In the post-communist world, the future of St Petersburg – renamed once more – seems more predictable than that of Moscow and it is now discovering a new identity as a centre of art and tourism, based on its imperial past. Its founder set out to create a European city on the classical model, and his achievement, after events he could never have imagined, has endured.

Left: The Stone Bridge, also known as All Saints Bridge, built over the Moscow river in the seventeenth century. Watercolour by F.A. Alexeiev, 1801, in the Kremlin State Museum.
The British Library, LB 31.c.7277

Below: St Petersburg in the mid eighteenth century: a view looking west along the Neva, showing the royal ships and the original bridge of boats.
The British Library, Maps K4 TAB 44

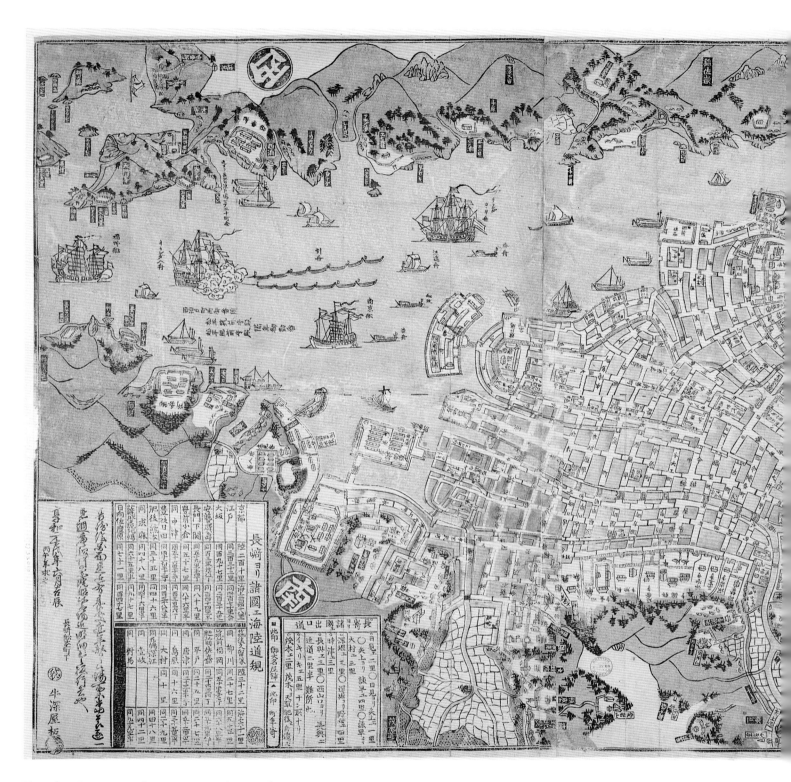

Nagasaki: a Japanese woodcut map, *c.*1680. The curved
shape of Deshima Island, the privileged Dutch trading
station, is clearly visible in the harbour, as are some large
Dutch ships, side by side with the Asian vessels.

Nagasaki

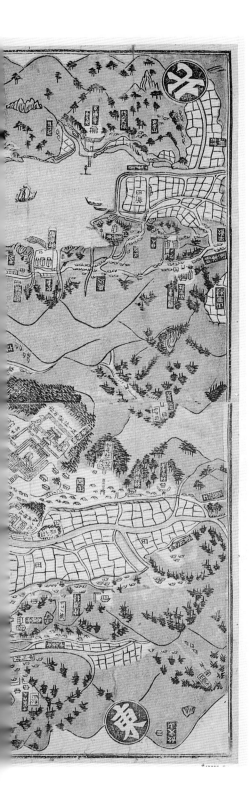

On 9 August 1945, three days after the bombing of Hiroshima, an American aircraft set out to drop the second atomic bomb on the Japanese city of Kokura. As dawn broke, however, Kokura was blanketed in dense cloud and the pilot altered course to the second target city of Nagasaki. In the explosion which followed, some 40,000 people were killed outright, and one third of the ancient city was obliterated. By such a slender chance was one city spared to live in obscurity, while another acquired a tragic fame as a symbol of destruction.

Nagasaki has the appearance of an amphitheatre, with its tiered houses rising high above the bay. It was a fishing village, recorded as early as the twelfth century, whose fortunes changed decisively around the year 1550, when its rulers opened it as a trading post to the European ships which had suddenly appeared in Japanese waters. With the explorers and traders came the Jesuit missionaries led by St Francis Xavier. They brought western science and scholarship to Japan, and they in turn sent back to Europe the first accounts of Japanese life and culture. Initially welcomed by the Japanese as bringers of new knowledge and technology, an anti-European and anti-Christian reaction soon set in and in 1597 a number of Christians, Japanese and foreign, were crucified on Nishizaka Hill, their deaths now commemorated in a monument on the site. In 1600 William Adams, an English shipwright serving aboard a Dutch ship, landed in Nagasaki and became the first Englishman to settle in Japan. He was employed by the Shogun as shipbuilder and as adviser on European affairs, and became so valued that he was never permitted to leave, living to become a semi-legendary figure in the orient.

Japanese suspicions of the European visitors eventually culminated in 1639 in the policy of *sakoku* – seclusion – which cut Japan off from all contact with the outside world. Christianity was proscribed, Europeans were expelled, and the Japanese were forbidden to travel abroad. Nevertheless the Japanese realised the usefulness of keeping open a channel of communication with the wider world, and for this purpose they chose the Dutch, who had never engaged in missionary activity and whose sole interest lay in trade. A Dutch community was established on the tiny island of Deshima in Nagasaki harbour, which was linked to the mainland by a bridge. Here the officers of the Dutch East India Company lived and worked for two centuries as the only Europeans permitted in Japan. Once each year a small group of them was licensed to leave Nagasaki and travel to the capital, Edo, to pay homage to the Shogun. Nagasaki thus became the unique point of contact between Japan and Europe, where techniques such as printing, map-making and firearms manufacture were received by the Japanese. Japan's own culture was still jealously guarded however: in 1826 the leading astronomer Takahashi befriended the scientist Franz von Siebold, and made him a gift of maps of Japan, for which he was flung into prison, where he died, while von Siebold took back to Europe the most detailed maps of Japan ever seen.

When the Americans forcibly opened Japan to foreign trade in 1854, Nagasaki's unique importance declined a little, although it remained a major port, and indeed benefited from the new policy. Western-style shipyards were built and managed by Mitsubishi, and until 1903 part of the harbour was leased to the Russians as the winter port of their navy. Like Hiroshima, Nagasaki has recovered, been rebuilt and become a centre for anti-war movements. A few nineteenth-century buildings somehow survived the blast, including the house built by Thomas Glover, the British merchant who settled in the city, and which was reputedly the setting which inspired Puccini's *Madame Butterfly*. Deshima has once again been linked to the mainland, and a small museum commemorates the years when this tiny enclave served as a unique interface between Japan and the European world.

Naples, 1464: an exceptional early Renaissance painting in which the city is the principal subject, and not merely a backdrop. It was painted to celebrate the return of the galleys which had liberated the island of Ischia from French rule. This is the Naples of Alfonso V, the Spanish monarch who made the city his capital and in whose reign it enjoyed its golden age. His fortress, the Castel Nuovo, is prominent in the foreground.

Naples

The perception of Naples as a place of flawed or corrupted beauty is a persistent one. Its setting on the steep coast above the Tyrrhenian Sea must be the most spectacular of any large city in Europe. The classical city, founded as a Greek colony, was laid out in the area north of where the Castel Nuovo now stands, while private villas were built on the higher ground, especially to the west, on the Posillipo. The Romans, including Caesar, Caligula and Nero, built their villas along this enticing coast, in spite of the threatening presence of Vesuvius, which destroyed the towns of Pompeii and Herculaneum in 61 AD.

In the post-classical era, Muslim raiders in the eighth century devastated the region, and from the eleventh century onwards, Naples was the capital of the foreign dynasties which ruled southern Italy: Norman, Hohenstaufen, Angevin, Aragonese, Hapsburg and Bourbon. To the non-specialist historian, the sequence of events is almost incomprehensible. They brought their culture with them, but their legacy to the Neapolitans was one of political oppression and poverty. Periodic attempts at resistance ended inevitably in a welter of blood. The mystic poet and philosopher, Tomasso Campanella, was imprisoned in the Castel Nuovo for twenty-seven years and tortured for his part in an anti-Spanish resistance movement. This Spanish domination partly explains why Naples lacks the wealth of Renaissance architecture expected from an Italian city, although the catastrophic war damage of 1944 also has much to answer for. To counterbalance this, the Archaeological Museum contains the unique legacy of Pompeii and the great Farnese collection of antiquities. The Spanish quarter, the archetypal Naples slum which is actually a grid of steep, narrow streets

Ma.ª Scarino

Lo Scutillo

Sta Caterina

Royal Palace of Capodimonte

HILL OF SCUTILLO

HILL OF CAPODIMONTE

Ma della Stella

Ma Talamo

Ancient Sepulcre

Ma Villa

Ma Zua

il Celso

Observatory

Ma Tuorno

Conocchia

Li Canzani

Ma Certosini

Ma Morna

P. Vecchio

Dueporte

HILL OF DUEPORTE

le Fontanelle

Mt.e Pastore

Ma di S. Severo

Ma Romeo

Monte

delle

Donzelle

Infrascata

la Concezione

Arenella

Bottone

L'Architiello

Antignano

Loguto

Bottone

CASTEL S.t ELMO

Rocca to Soccavo

Valley of Soccavo

Mongibello

la Torre Sopra Soccavo

S. M.della Libera

il Vomero

Villa Floridiana

HILL OF MALL

Cas. del Duca della Regina

Villa Belvedere

Villa Ricciardi

S.t Franc.o di Paolo

l'Eremo

HILL OF CHIAJA

Cast di S.M. in Portico

Largo del Vasto

CASTELLO NUOVO

Villa Patrizi

Wet Dock

Piazza del Pri Reale

Arsenal

Chiaja

Vico freddo

Chiaja

Villa reale

Public parterre

Largo della Vittoria

Acqua Solforea

Grotta of Pozzuoli

Boschetto

Mergellina

HILL OF POSILIPO

Virgils Tomb

CASTELLO DELL' OVO

NOTE.

St. Strada	Street.	P. Ponte	Bridge.
Ca. Castello	Castle.	Ta. Taverna	Tavern.
Pi. Piazza	Place.	Po. Porta	Gate.
Pal. Palazzo	Palace.	M. Monasterio	Convent.
Te. Teatro	Theatre.	Ma. Masseria	Farmhouse.
† Church or Chapel.	Cas. Casino	Country house.
S. M.	S.ta Maria.

W. B. Clarke Arch.t dir.t

Castello nuovo

Castello S.t Elmo

Published by Baldwin & Cradock March 1835

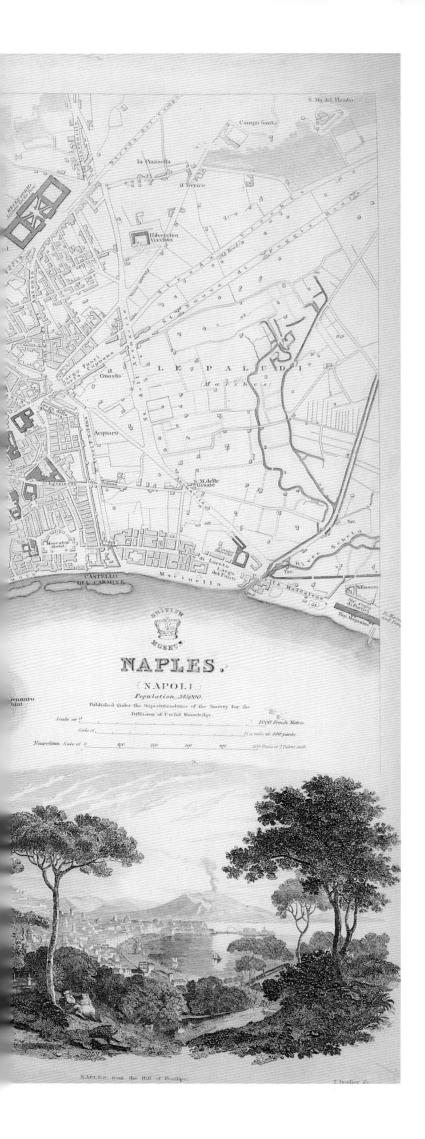

west of the via Toledo, dates from the late sixteenth century. Plague and earthquakes have periodically ravaged the city, and in 1884 an outbreak of cholera reputedly killed 1,000 people in a single night. This event galvanised the Italian government to take special action to heal some of Naples's wounds. Many slums were demolished and new aqueducts built to bring pure water from the hills.

In spite of its ambiguous reputation, Naples in the eighteenth, and still more the nineteenth, century became a fashionable destination for foreign visitors, drawn by the beauty of its setting to settle in the handsome villas of the Posillipo. The city and the region possessed an aura of hedonism, a dreamy paganism which appealed to visitors from northern Europe. This reputation is promoted in the novel *The Lost Stradivarius* (1895) by J. Meade Falkner, where Naples is the scene of an Englishman's spiritual collapse under the spell of pagan mystery rites once practised in caves beneath his villa. In the early twentieth century, the world's attention was captured by stories of the *Camorra*, the Neapolitan mafia, which, combined with the legacy of 1930s depression and the terrible damage inflicted on the city in World War II, have all left Naples with the reputation of an orphan among Italian cities, sullied by crime and poverty, an image only now being rectified. The past exists in Naples, but in a way which is unwelcome and haunting compared with the Renaissance cities of the north.

Naples in 1848: the functional street plan is complemented by the classic view over the city from the hill of Posillipo towards Vesuvius.
The British Library, Maps 38.e.8

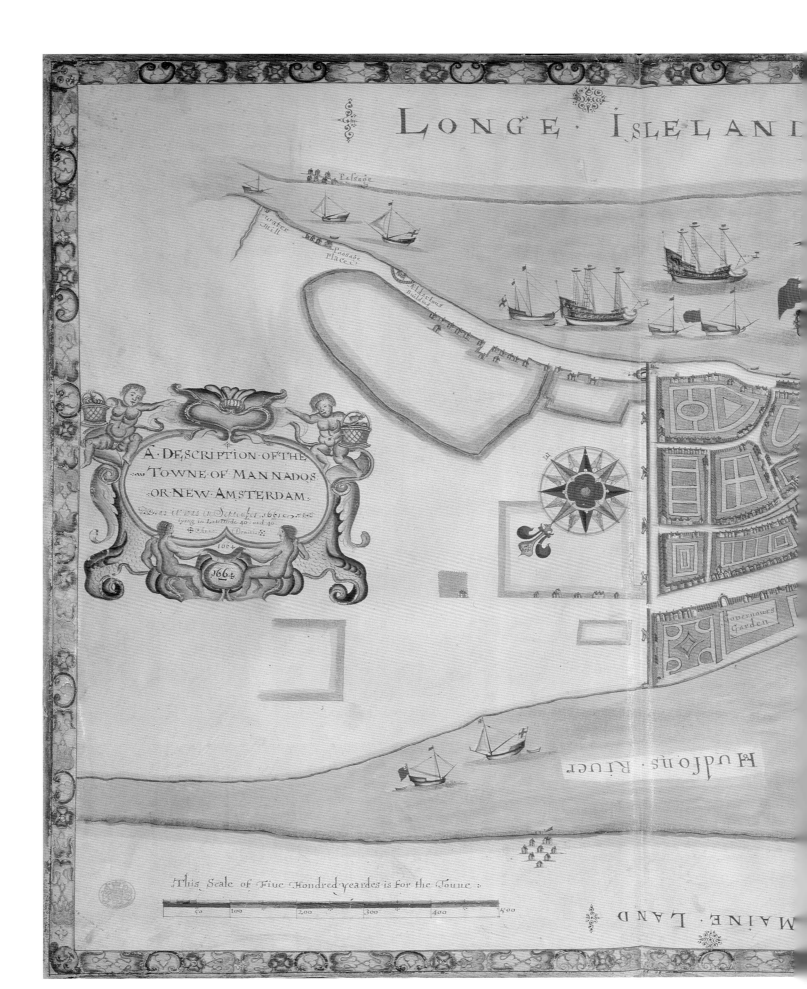

LONGE · ISLELAND

Palsade

Graber
Mill

Passage
Place

A Description of the
Towne of Mannados
or New Amsterdam.
As it was in September 1661
lying in Latitude 40 and 40
Anno Domini
1664
1664

Governours
Garden

Hudsons River

This Scale of Five Hundred yeardes is for the Towne

50   100   200   300   400   500

MAINE · LAND

New York

The earliest origins of New York City were fittingly international: the first European to sail into the bay in 1524 was an Italian in the service of the king of France – Giovanni da Verrazano, after whom the narrows are named. A further eighty-five years were to pass before a second and closer look was taken in 1609, and this time it was an Englishman commissioned by a Dutch exploration company to seek a way through or around the American continent to the Pacific. Henry Hudson reported enthusiastically on this magnificent harbour, the sheltering green hills and the temperate climate. Within a few years a small Dutch colony had been established, and in 1626 the governor, Peter Minuit, bought the territory of Manhattan for sixty guilders' worth of merchandise – around twenty-five dollars. New Amsterdam was declared a city in 1653, when its population stood at around 800. But European rivalries meant that there was to be no Dutch Empire in the New World: an English fleet seized the city in 1664, and the name was changed to New York, in honour of the Duke of York, brother to King Charles II.

These events form the background to the colourful map of New York known as 'The Duke's Plan'. The map is anonymous and its title is enigmatic: A Description of the Towne of Mannados or New Amsterdam, as it was in 1661, and a second date, 1664, is added below. The ships surrounding Manhattan are ostentatiously flying British flags, and the inference is that this is a British copy, made in 1664, of a Dutch plan drawn three years earlier. The city is confined here to the original Manhattan settlement, bounded by the wall where Wall Street would be built. Beside the fortification where Battery Park now stands is a picturesque legacy from the Dutch: a windmill. These were eventful years, for in 1673 the Dutch reclaimed their colony in a seaborne raid, and the city became briefly 'New Orange'. An Anglo-Dutch peace treaty of the following year ended this game of chess, and British rule was confirmed. In the following hundred years Manhattan lived by trade, including the spoils of piracy for which it was notorious, while across the river farms and estates prospered in New Jersey and 'Brookland'.

A century after the Duke's Plan was drawn, Bernard Ratzer, a military surveyor, drew up a painstaking map of New York which was published

New York, or 'New Amsterdam, as it was in 1661', as stated in an anonymous manuscript map dating from three years before the English took and renamed the town. The line of the future Wall Street and Battery Park are both clearly visible.
The British Library, Maps K. Top. CXXI.35

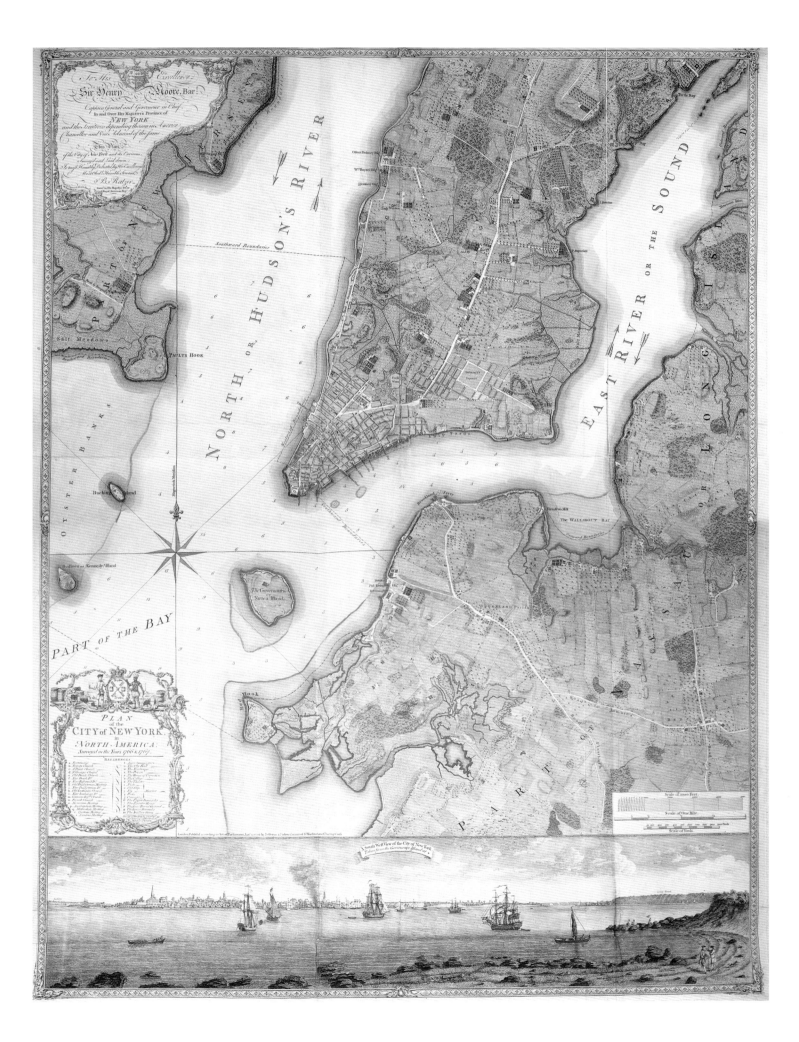

in 1776, just when the War of Independence was gathering force. This time Manhattan is mapped northwards along Bowery Lane as far as a line running along 57th Street and the southern tip of Roosevelt Island. At the foot of the map is a distant panoramic view of the buildings of Manhattan seen from Governor's Island. On 'Bedloes or Kennedy Island' the Statue of Liberty now stands, while 'Bucking Island' is now Ellis Island. A number of individual landowners are named in their lands, among them the Stuyvesants' Estate facing the East River. By the time this serene-looking map had been engraved, fighting had already broken out between patriot groups and British soldiers, and huge fires had destroyed parts of the city. After a decade of historic conflict, New York emerged as the first capital of the new nation, and the scene of the inauguration of Washington as the first president. The population passed 60,000 as the nineteenth century opened.

What was it that transformed this compact city into the giant metropolis of the twentieth century? Undoubtedly it was the twin forces of immigration and industrialisation. For the millions who arrived from Europe, New York was the gateway city to America, but many chose to stay and to man the new industries. The tide of settlement engulfed huge areas to the east and the west, but the heart of New York has always been the Manhattan peninsula which so impressed Hudson, and which Peter Minuit acquired for those twenty-five dollars – surely the greatest bargain in recorded history.

Opposite: New York as surveyed in 1766 and published in 1776, by Bernard Ratzer. Originally drawn in 1770 by a British survey officer, this map was reissued in the early days of the War of Independence, and just weeks before much of the city was destroyed by fire.
The British Library, Maps 1 TAB 44 (28)

Below: The lower Manhattan waterfront in 1884, a dark industrial seaport scene, not essentially different from that of Liverpool or Hamburg at that time.
The British Library, Newspaper Library, Colindale
x702/5596 243

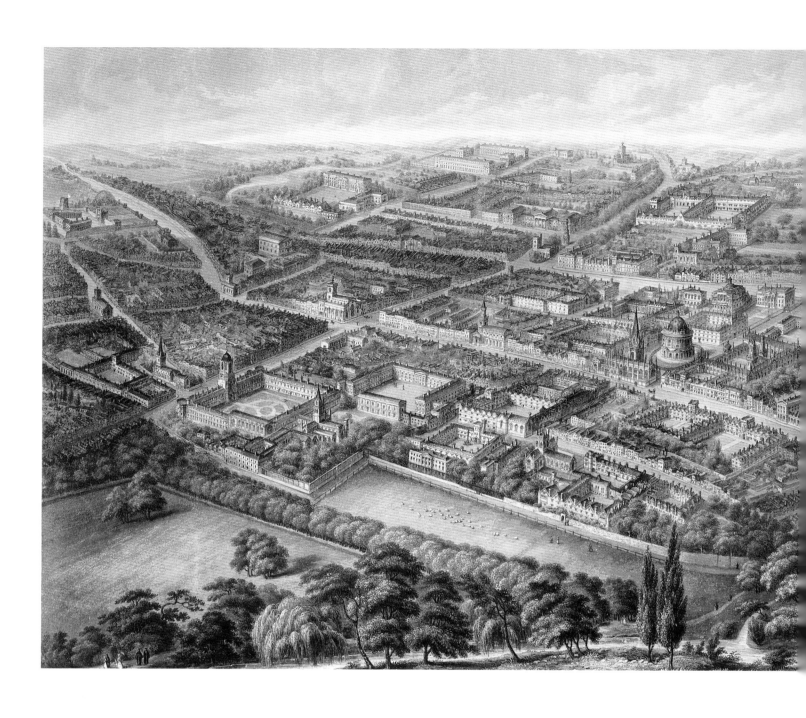

Oxford

Above: Oxford, 1848, by Nathaniel Whittock. A superb view of nineteenth-century Oxford, as if seen from some non-existent hilltop east of the Cherwell, above the Iffley Road. All the colleges and other landmarks - the Bodleian, the Ashmolean, the University Press - are shown in meticulous detail. This beautiful engraving has a serenity - perhaps caused by the almost total absence of people - which makes one long for the Oxford before the days of tourist crowds, shopping centres and traffic jams.
The British Library, Maps 4735 (8)

Left: The Clarendon Building, Oxford, designed by Hawksmoor and erected 1711-13. It was the original home of the Oxford University Press and is now part of the Bodleian Library. From Nicolas Barker's *The Oxford University Press and the Spread of Learning*, 1978.
The British Library, 2708.ff.39

The romantic conception of Oxford, as a city where the thought and the poetry of the past are somehow felt to be tangible in the air and in the stones, is well evoked in Compton Mackenzie's *Sinister Street* of 1914:

Nothing ever quite equalled those damp November dusks, when after a long walk through silent country Michael and Alan came back to the din of Carfax and splashed their way along the crowded and greasy Cornmarket towards St Giles, those damp November dusks when they would find the tea-things glimmering in the twilight. Buttered toast was eaten, tea was drunk; the second-best pipe of the day was smoked to idle cracklings of The Oxford Review *and* The Star; *a stout landlady cleared away, and during the temporary disturbance Michael pulled back the blinds and watched the darkness and fog slowly blotting out St John's and the alley of elm-trees opposite, and giving to the Martyr's Memorial and even to Balliol a gothic and significant mystery. The room was quiet again; the lamps and the fire glowed; Michael and Alan, settled in deep chairs, read their history and their philosophy; outside in the November night footsteps went by; carts and wagons occasionally rattled; bells chimed; outside in the November murk present life was manifesting its continuity; here within, the battles and the glories, the thoughts, the theories and the speculations of the past moved for Michael and Alan across the printed pages under the rich lamplight.*

After dinner Michael and Alan read on towards eleven o'clock, at which hour Alan usually went bed. It was after his departure that in a way Michael enjoyed the night most. The medieval chronicles were put back on their shelf; Stubbs or Lingard, Froude, Freeman, Guizot, Lavisse, or Gregorovius were put back; round the warm and silent room Michael wandered uncertain for a while; and at the end of five minutes down came Don Quixote or Adlington's Apuleius, or Florio's Montaigne, or Lucian's True History. The fire crumbled away to ashes and powder; the fog stole into the room; outside was now nothing but the chimes at their measured intervals, nothing but the noise of them to say a city was there; at that hour Oxford was truly austere, for it was neither in time nor in space, but the abstraction of a city.

It is this romantic notion of Oxford that draws visitors in their thousands, some serious and scholarly, others hurried tourists. There can't be many places in the world where learning and intellectual achievement have become tourist attractions, but Oxford is one of them. Do people find what they are looking for? Can that community with the past survive in this age of mobile phones, fast food and traffic jams? It seems doubtful, and Mackenzie's classic evocation of a vanished Oxford provides a link with the past that is as precious and permanent as the domes and spires of the colleges themselves.

1593 PASCALE CICONIA DVCE VENETIAE FECE AN DNV

Noua PALMAE ciuitas in
patria Foroiuliensi ad maris Adria
tici ostium contra Barbarorum in
cursum à Venetis ædificata

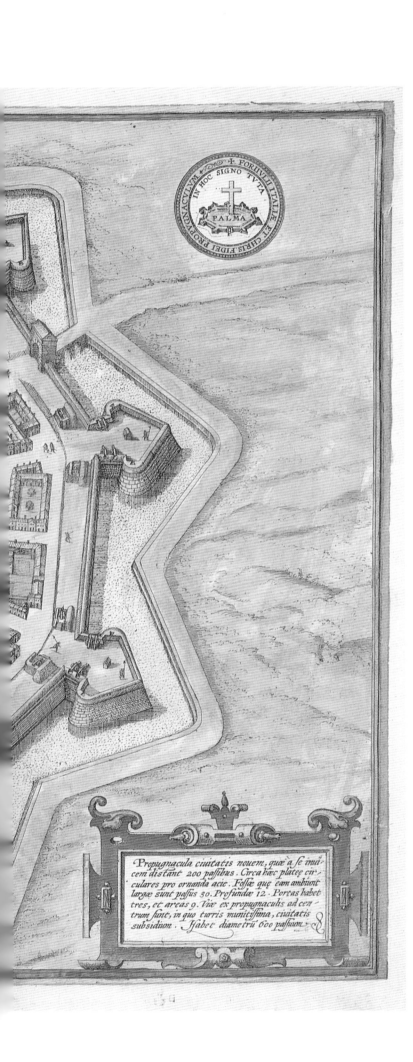

Palmanova

Constructed in the decade between 1593 and 1603, Palmanova brings together two principles of Renaissance city building: the needs of military engineering, and the geometrical regularity of the ideal city. Social amenities played no part in the merging of these two themes. Lying halfway between Udine and the sea, Palmonova was built by the Venetians as a defensive post against the Austrians, and against possible Turkish landings. It was planned as a complete whole, with no change of shape or purpose envisaged in the future. From the central tower, six roads radiate to the outworks of the fortifications, every other road leading to a gateway. From this central watchtower, the progress of any siege could be monitored and reinforcements directed. The radial avenues are intersected by three concentric ring roads. In those roads which do not lead to a gate, open squares have been formed. The whole effect is one of perfect symmetry. The massive strength of the outer walls can be gauged from the difference in the inner and outer diameters: within the ramparts it is 2,400 feet (730 metres), while outside the bastions and moats it is 5,400 feet (1,650 metres). The city was designed by the Vicenza-born architect Vincenzo Scamozzi (1552–1616), theorist of late Renaissance architecture, disciple of Palladio, and one of the intellectual fathers of neoclassicism. This form of fortification, and the rage for geometric regularity, were both hugely influential, and were imitated wherever possible by cities throughout Europe.

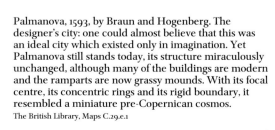

Palmanova, 1593, by Braun and Hogenberg. The designer's city: one could almost believe that this was an ideal city which existed only in imagination. Yet Palmanova still stands today, its structure miraculously unchanged, although many of the buildings are modern and the ramparts are now grassy mounds. With its focal centre, its concentric rings and its rigid boundary, it resembled a miniature pre-Copernican cosmos.
The British Library, Maps C.29.e.1

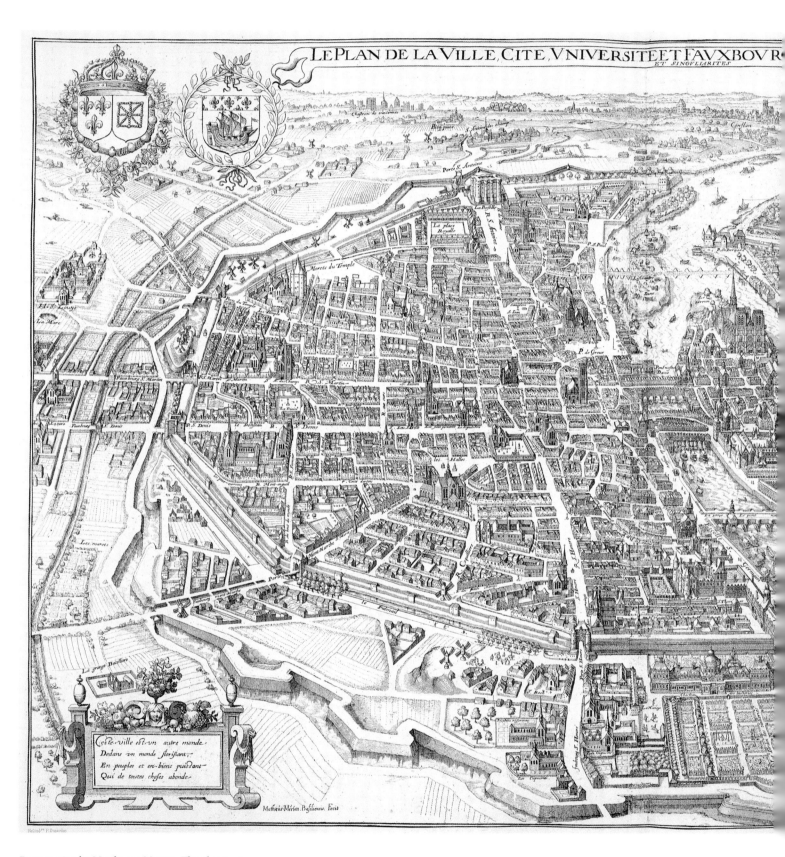

LE PLAN DE LA VILLE, CITE, VNIVERSITE ET FAVXBOVR
ET SINGVLIARITES

Ceste ville est vn autre monde.
Dedans vn monde florissant,
En peuples et en biens puissant,
Qui de toutes choses abonde.

Mathæus Merian Basiliensis Fecit

Paris in 1650 by Matthaeus Merian. The classic view
from the west with the old fortresses, the Louvre and
the Bastille, on the left, and the Isle de la Cité and
Notre Dame as the central focus.
The British Library, Maps 152.d. 4xv

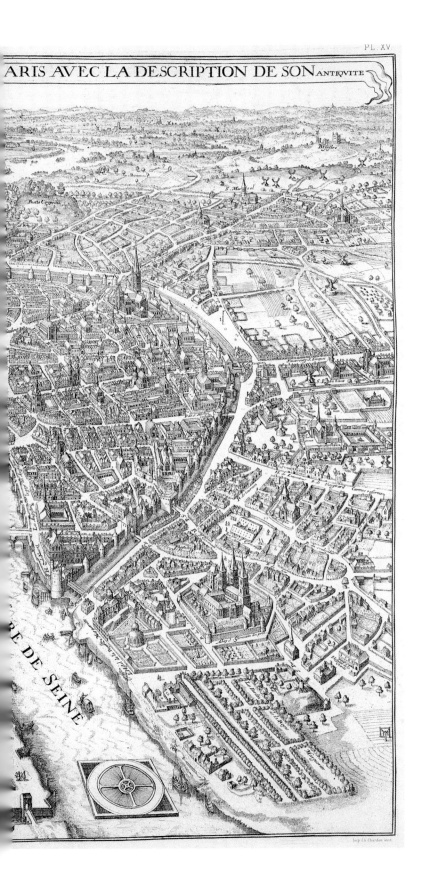

Paris

Paris is one of those capital cities whose story is so bound up with the destiny and consciousness of a nation that their two histories are inseparably merged together. Despotic monarchy, enlightenment, revolution, empire, industrialisation, artistic brilliance, intellectual arrogance – all the major themes of French history have been focused in the public and private spaces of Paris, and street violence has long been part of that history.

The Roman town described by Caesar was concentrated entirely on the Isle de la Cité, and was inhabited by a tribe called the Parisii. In the fifth century, when the Franks migrated west from the Rhineland, their king, Clovis, adopted Paris as his capital and settlement spread to both banks of the Seine. The medieval city prospered under the Capetian kings and became divided into three recognised areas: the Isle, the centre of church and government; the right bank, the commercial quarter around the Hôtel de Ville and the markets; and the left bank, home to the university of the Sorbonne, founded around the year 1200, and where the first French printing press would operate in 1469. Two great fortresses guarded the city to the north of the Seine, the Louvre in the west and the Bastille in the east, with the city walls looped between them. A few miles to the north, the abbey of St Denis became the mausoleum of the French monarchs. Always threatened by civil unrest and conflict with England – the city was under English rule from 1420 to 1436 – the kings normally resided not in Paris but in Touraine. It was with the reign of Francis I and his successors, from the 1520s onwards, that Paris emerged as a royal capital with a Renaissance court. The Louvre was rebuilt as a palace, and the country to the west became a royal domain. The Tuileries and the Luxembourg palaces followed, as did the Pont-Neuf, the first bridge over the Seine other than those to and from the Isle. The wars of religion from 1560 to 1600 brought the French state close to collapse, and the great prize was Paris, and who was to rule there.

With the reign of Louis XIV, the history of Paris entered a new phase. Although he moved his residence to Versailles, 10 miles (16 km) from the Louvre, the king was determined to make Paris reflect his cult of personal glory. His minister of works, Colbert, agreed that, next to winning battles, building on a grand scale was most calculated to enhance the king's prestige. The Tuileries and the Louvre were rebuilt, the Invalides founded, the tree-lined avenue of the Champs Elysées laid out for fashionable carriage-driving, complemented across the city by the Parc de Vincennes, and the rues St Honoré and St Germain were lined with the elegant houses of courtiers.

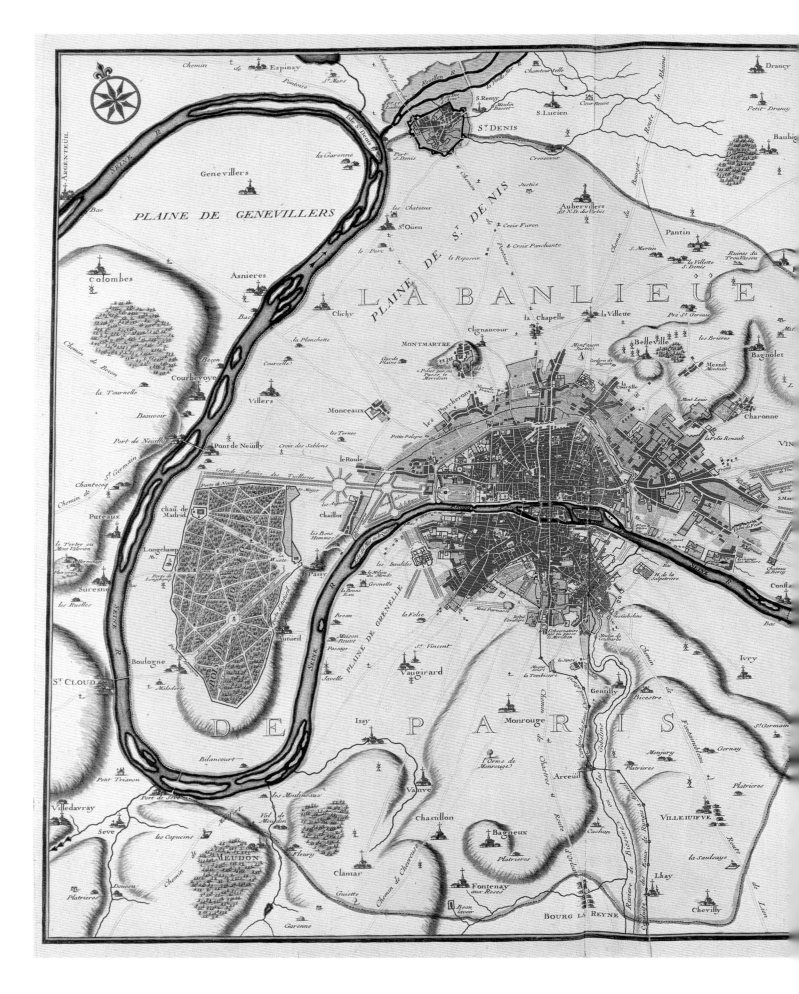

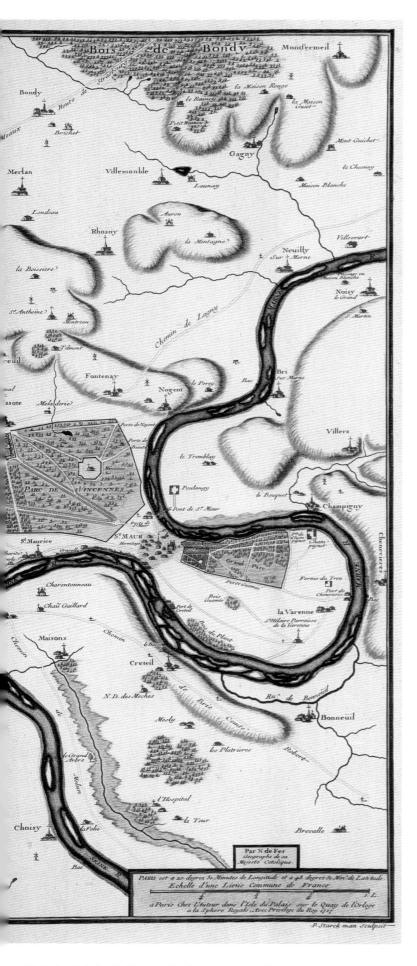

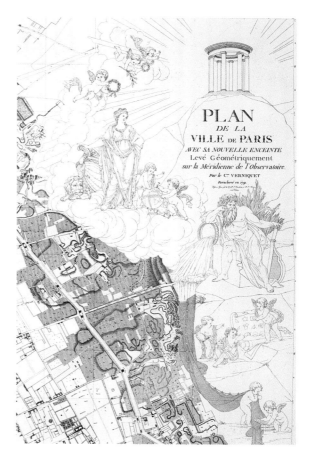

Paris, by Nicholas de Fer, 1717. The fascination with this plan is to trace all the landmarks which then lay well outside the city centre, and which have now been engulfed by it: the sites of Sacré Coeur, Père Lachaise, the Bois de Boulogne, and many others.
Bodleian Library

Above (both): Paris: two contrasting map cartouches, the first typifying the taste of the *ancien regime*, the second of the revolutionary era.
The British Library. Maps 152.d.4 XXV and XXXI

Baroque Paris fulfilled: Louis XIV visiting Les Invalides,
the magnificent church which epitomised the Sun King's
imprint on his capital. Painting by Pierre-Denis Martin.

Musée de la Ville de Paris, Musée Carnavalet, Paris/www.bridgeman.co.uk

This new royal city breached the old city walls to the west, but to the north and east at the wall of the 'tax-farmers', tolls were still levied for entry through the gates. This was the city depicted in the 1717 plan by de Fer, and in the extraordinarily detailed panoramic view of 1739 known as the Turgot plan. It was the city of elegance, of royal palaces and vistas, of enlightenment.

But behind the façade, the city was seething with a quarter of a million discontented citizens, whose miserable dwellings did not adorn the Turgot plan. Even at the commencement of Louis XIV's reign, Corneille had mocked:

> *'This whole city, built in splendour*
> *Raised by miracle from an old ditch,*
> *Would trick us into believing that all its people*
> *Are gods or kings.'*

Voltaire too had warned that Paris was full of destructive vortices, and even at the height of the age of enlightenment, society's victims were still publicly tortured and torn to pieces before the Hôtel de Ville.

The Revolution, the greatest drama in French history, took Paris as its stage – the Bastille, the Place de la Révolution (now Concorde), Versailles, and the innumerable Jacobin clubs throughout the city. The cathedral of Notre Dame narrowly escaped destruction, while a newly built church was hastily dedicated as The Pantheon, a secular temple to the greatness of man. Who knows what the face of Paris might have become, had not Napoleon revived the royal theme of glorification in his rule over the capital: triumphal arches, columns of victory and new bridges were his legacy. Strangely, however, the streets had still not been fundamentally reshaped since the Middle Ages, and by the middle of the nineteenth century several forces came together to hasten the overdue change. The population had passed one million, and the huge increase in carriage traffic was choking the centre; the coming of the railways had produced new focal points in the city; the revolution of 1848 had shown a shaken government how easily the narrow streets could be barricaded and held by a mob; and finally the emperor Napoleon III wished to make Paris a showpiece capital worthy of his grandiose ambitions. He commissioned Baron Georges Haussmann, Prefect of the Seine, to carry out his scheme, and between 1855 and 1870 Haussmann demolished slums, created wide boulevards and squares, and laid out public parks. He has been credited with shaping modern Paris, but he has also been criticised for pandering to the French taste for self-aggrandisement, for ornate heartlessness.

That Haussmann's Paris was a veneer concealing corrosive social divisions seemed to be swiftly proved. In an uncanny mirror image of the crisis of eighty years earlier, Paris was traumatised again by war and revolution. The Franco-Prussian War ended in humiliation in 1871: the seat of government was transferred to Bordeaux and, after the military collapse, the French were compelled to witness a German triumphal procession through the Arc de Triomphe. The Paris Commune rebelled against the new republic, and was ferociously suppressed with thousands of deaths. It seems almost incredible that the city recovered within a few years to become once again the host city of international exhibitions and the Eiffel Tower, of the Impressionists and of Montmartre, and all the symbols which evoke *la belle époque*. The conscious search for grandeur has always motivated the personalities who have shaped Paris, and their monuments have bequeathed us a unique city, but one with an unmistakably darker side, which has erupted again and again throughout its history.

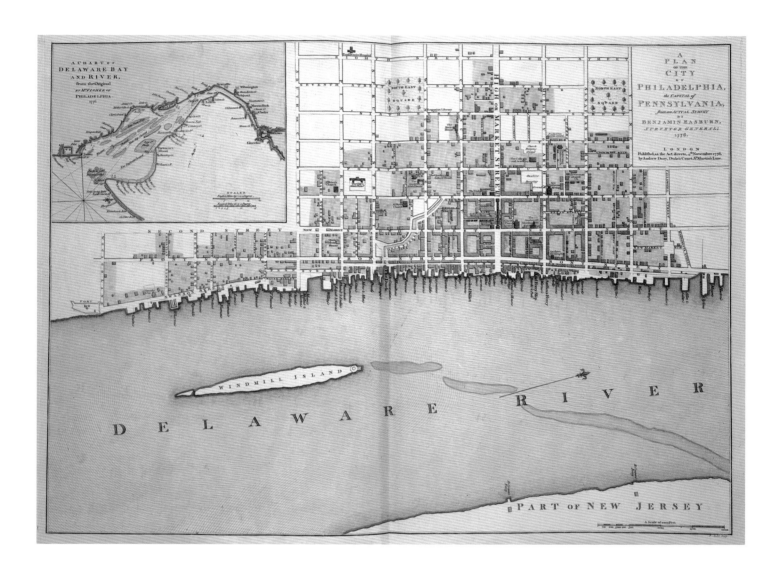

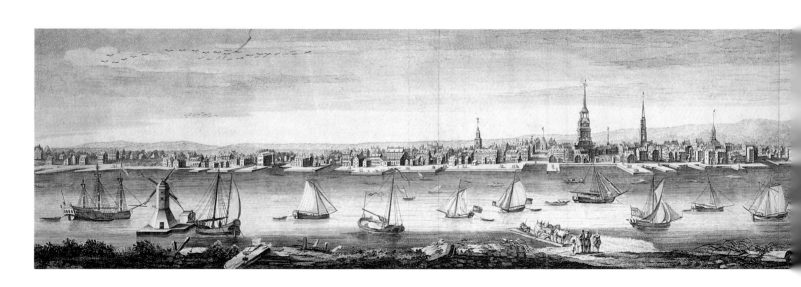

Philadelphia

William Penn's 'holy experiment', a colony embodying political and religious toleration, involved also an experiment in urban planning which was one of the most successful in the New World, and which deeply influenced other American cities. Penn was a Londoner who had lived through the plague and the fire of 1665 and 1666, and was familiar with the splendours and miseries of the old European city, and it seems clear that the founding of a new city was central to Penn's vision.

The opportunity to realise that vision came through a large debt of £16,000 owed by King Charles II to Penn's father, Admiral Sir William Penn. In 1681, in return for the cancellation of this debt, the king was persuaded to grant the younger Penn a charter to an extensive new colony to be called Pennsylvania, where Penn planned to settle his fellow Quakers, and so free them from persecution at home. Those who bought land in the colony were to be allotted a proportionate site in the new capital. By the summer of the following year, Penn's surveyor, Captain Thomas Holme, was at work on the banks of the Delaware River close to its junction with the Schuylkil, laying out a 10,000-acre site on which to build the city, whose name Penn had already chosen and which signified 'brotherly love'. Penn was most concerned to secure the friendship of the Indians, and had his surveyor negotiate a treaty with them at the outset, which Voltaire admiringly called 'the only treaty not sworn to and never broken'. Penn was the guiding spirit behind the design, which was to be of wide, uniform street blocks:

'Be sure to settle the figure of the town so as that the streets hereafter may be uniform down to the water from the country bounds ... let the house be built upon a line ... let every house be placed in the middle of its plat, so that there may be ground on each side for gardens or orchards or fields, that it may be a green country town, which will never be burnt, and always be wholesome.'

This plan was embodied in a description of the city which was published in London in 1683, and which clearly shows the central square and the four smaller squares placed symmetrically around it. In London, Penn had lived in Lincoln's Inn with its adjoining open fields, which, unlike London's other squares, were not reserved for the use of wealthy residents, and the squares in Philadelphia were to be vital public spaces in the new city. The clearest influence upon Penn's design were some of the plans produced (but not actually realised) for the rebuilding of London after the fire of 1666, especially that of Richard Newcourt, which exhibited precisely this pattern of five squares and a grid of regularly spaced streets. Penn and Holme would, like all other educated Londoners, have seen and discussed these plans.

Penn was called back to England early in 1684, but by that time the city had advanced rapidly to more than 300 houses, 'large and well-built with good cellars, three stories, and some with balconies ... there is also a fair key about three hundred foot square, to which a ship of five hundred tuns may be laid here broadside.' Such was the success of the new community that Penn estimated that the worst lot in the new town was already worth four times more than when it was laid out, while the best was worth forty times. This rise in land values led inevitably to the dilution of the strict city plan, with additional narrow streets being cut into the original ones, so that more small houses could be packed in.

Economically the town was thriving. Pennsylvania's tolerant religious policy made it attractive to immigrants from many countries and backgrounds, while the rivers gave access to farming and mine products from the hinterland, and allowed onward shipping to Europe. By 1770 the population was 30,000, the coal and iron industries in the west of the state were established, and Philadelphia was reputedly the third business centre of the British Empire, behind only London and Liverpool. It was natural therefore that the city should be at the forefront of the events leading to revolution and independence. Merchants and political philosophers participated in the congresses in which the identity of the new nation took shape. After intense debate spread over twenty-two months from September 1774 to July 1776, much of it taking place in the State House, the Declaration of Independence was written and proclaimed. The city raised five battalions for the inevitable war, but was occupied almost at once by British troops following the American defeat at the Battle of Brandywine. After almost a decade of war and uncertainty, the Constitution of the United States was hammered out in the same building, and for ten years Philadelphia was the capital of the United States, during the planning of Washington. As the site of Penn's unique experiment in colonial idealism, and as the birthplace of the United States, the historic centre of Philadelphia has a unique place in America's urban history. Other towns in Pennsylvania - Lancaster, Reading and Pittsburgh - copied the square grid design interspersed with open squares, while the same concept was used further afield in Raleigh, North Carolina and Tallahassee, Florida.

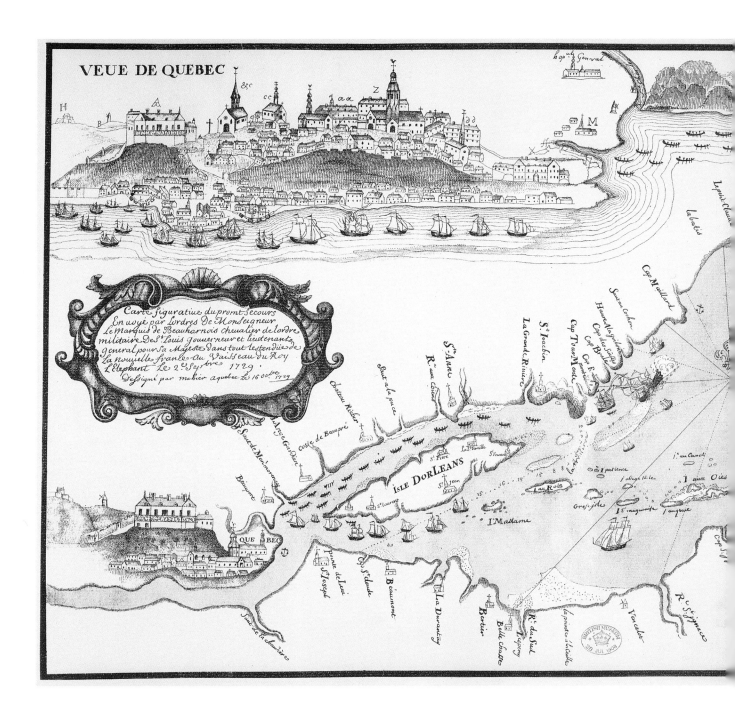

VEUE DE QUEBEC

Carte figuratiue du promt Secours
Enuoyé par l'ordres De Monseigneur
Le Marquis de Beauharnois cheualier de l'ordre
militaire De St Louis gouuerneur et lieutenant
general pour Sa Majesté dans tout l'estendüe de
La nouuelle Franse Au Vaisseau du Roy
L'Elephant Le 2e Septbre 1729.
Dessigné par mahier a quebec Le 16 Octobre 1729

LE DE CINQ LIEUES·

Above: Quebec, showing its geographical setting on a natural citadel of rock overlooking the point where the St Lawrence narrows drastically.
The British Library, Maps 115.d.28 (7)

Left: Quebec, 1729, the only counterpart in North America of the European hilltop fortress town, with its citadel, cathedral and defensive walls, all built in an imposing architectural style suited to the capital of New France.
The British Library, Maps 115.d.28 (11)

Quebec and New Orleans
the cities of New France

Throughout the seventeenth and eighteenth centuries, a vast swathe of territory in North America was known to Europeans as 'New France', and its limits were defined by two historic settlements – in the north was Quebec, and in the south was New Orleans. The earlier foundation by more than a hundred years, Quebec is the oldest non-Spanish city in North America, and is now the only surviving walled city in the continent. The cradle of French Canada, it occupies a strong strategic position on a rocky promontory at the point where the broad estuary of the St Lawrence River suddenly narrows to little more than half a mile in breadth. The site was perfectly chosen to supply and control trade and communication with the interior of the continent. It was regarded as the point of entry to a vast but tenuous French empire, which imagination spread throughout the Great Lakes region and south along the Mississippi.

The first European to visit the site of Quebec was Jacques Cartier in 1535. He found there an Indian village called Stadacona, and returned again in 1541 with a more determined colonial party. However, the bitter winter and the failure to find the gold which European explorers inevitably sought discouraged further French interest in Canada for many years. Much later in the sixteenth century they began to return, drawn by the cod fisheries off Newfoundland, and by another valuable commodity which abounded there – furs. The French crown granted fur-trading monopolies to nobles who would finance exploration and settlement, and in the summer of 1608 Quebec was chosen as the headquarters of such an expedition, led by Sieur de Monts but masterminded by Samuel de Champlain. Champlain was an expert geographer and cartographer who had already spent time in Spanish America, and had taken careful note of such towns as Mexico City, Santa Domingo and Vera Cruz. The first buildings in Quebec were of course of wood, in the lower district by the water's edge, but Champlain soon realised that a stronger fortified position should be erected on the cliff-tops, which rose some 300 feet (90 metres) above the river. Thus two distinct areas began to evolve in the new city, similar to those of many European cities: the *Haute Ville*, centred around the fortress, and the *Basse Ville* lying below or around it. Political and religious institutions were sited in the upper town, commerce and labour down around the harbour. The Jesuits arrived in 1635 and established a college, and in 1674 Françcois de Laval was appointed the first bishop of Quebec, the seminary which he founded forming the nucleus of Laval University.

The city grew rapidly after 1660, under royal direction from France. Churches, residences and hospitals were built of stone, and the walls around the upper town were completed by 1729. Fishing, ship-building and fur-trading flourished in the harbour, and despite the growth of Montreal, 100 miles (160 km) upstream, Quebec acted as the capital of New France, from where trappers, farmers and missionaries set out into the interior. By 1750 Quebec presented the appearance of a typical European town, with its citadel and cathedral sited on high ground above a busy trading river, and with a population approaching 10,000. The city was, at this point, on the eve of the most momentous change in its history. Quebec became a prize in the Anglo-French war of 1756–63, for it was rightly seen as the strategic gateway to New France. The story of the British seizure of Quebec in September 1759, in which both the English and the French commanders, Wolfe and Montcalm, lost their lives, is one of the landmarks of Canadian history, one whose significance has never ceased to echo down the years. Scarcely less significant was the failed attempt by American revolutionary forces to retake the city from Britain just sixteen years later: had this succeeded, the whole of Canada might eventually have become part of the United States. In the event, this siege of 1775-76, in which the lower town was taken but not the fortress, was the last warlike scene in Quebec's history, and the city was never attacked again. Economically it was steadily overtaken by Montreal, but its status as the most historic city in Canada was always acknowledged.

New Orleans, 1769. The archetypal planned colony, with
its chequerboard grid of streets imposed on the swamps
or forests of the New World, a pattern which survived to
become the heart of the future city. When this plan was
published, the city was under Spanish rule, having been
handed over by Louis XV in a secret treaty; only in 1803
did New Orleans become a city of the United States.
National Archives

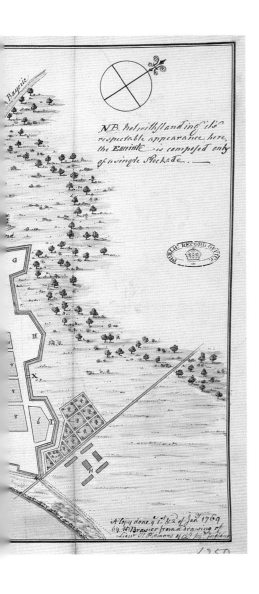

At the southern extremity of New France, two brothers, Pierre d'Iberville and Jean Baptiste de Bienville, made two expeditions from France in 1698 and 1701 to locate and explore the Mississippi River from the south, from the Gulf of Mexico. They were following up the 1681 expedition of Robert de La Salle, whose probing of the Mississippi from the Great Lakes to the sea was the basis of the French claim to that entire region. D'Iberville and de Bienville established the settlement which later became Mobile, Alabama, but de Bienville still favoured a new city on the Mississippi itself. In 1718 he chose a crescent-shaped bend in the river, some 80 miles (130 km) inland from the sea, where an Indian portage gave access to Lake Portchartrain. Work to clear the site and erect the first buildings was assisted by the schemes of a disreputable Scottish financier, John Law, who set himself up in Paris and sold stock in a *Compagnie d'Occident*, describing a great city on the Mississippi which was rising from the swamps, and which was only slightly less magnificent than Paris itself. Three years later, Law's company had collapsed, but the engineers had already laid out the new town on the classic grid pattern which still survives in the *Vieux Carre*, and named it for the Regent of France, the Duc d'Orleans. The labour was great, for the land was swampy and fever-ridden, rising only a few feet above sea level. The water table was so low that the tradition began of burying the dead in raised tombs, for burial in the earth was impossible. Embankments or *levées* were raised against the river; later they would be planted with orange trees and become fashionable places for evening promenades. The focal point of the new city was the *Place d'Armes*, the modern Jackson Square, where the first St Louis Cathedral was erected by 1720; this was twice destroyed, by hurricane and by fire, the present cathedral being completed in 1794.

With its swampy location and lack of rich natural resources, the capital of Louisiana did not prove an attractive lure for French colonists, and the early inhabitants were soldiers, trappers, convicts and indigents, for it was occasionally the practice of Parisian courts to banish offenders to New Orleans, although it was never officially a penal colony. Plantations of tobacco and rice were established outside the town, and slaves were brought in to man them, but growth was precarious. In 1762 France decided to rid itself of the struggling colony: Louis XV transferred it to Spain in a secret treaty, which remained unknown to the inhabitants until the new Spanish governor actually arrived with his soldiers in the city. The Spanish ruled for a little over three decades, but left a considerable mark on the architecture. In 1800 Napoleon negotiated the return of the city and of the whole of Louisiana to France, but he soon relinquished any strategic ambitions in North America, and New Orleans formed part of the Louisiana Purchase of 1803. At last integrated with its natural American hinterland, the city began to thrive with exports of cotton, timber, liquor and finished goods. Only the war with Britain from 1812 to 1814 threatened its stability, and the celebrated Battle of New Orleans, when Andrew Jackson's force routed a seaborne British attack, famously took place two weeks after the war had actually ended, for news of the peace had not reached either side. This victory possibly saved Louisiana for the Union, and avenged the burning of Washington by the British; it is commemorated in the equestrian statue of Jackson before the cathedral. The years that followed were the golden age of building in New Orleans, which created a gracious but relaxed urban style. The old fortifications were pulled down and replaced by broad boulevards, Canal Street and North Rampart Street. As the city spread east and west, the curving river bank formed the basis for new areas of square-planned streets, which were carefully linked to the original central grid. The unforgettable street names for which New Orleans is famous - Frenchmen, Socrates, Good Children, Clio, and above all Desire, with its echoes of Tennessee Williams - date from this period. The French contribution to urban America is small for historical reasons, but in Quebec and New Orleans, it has left two unique memorials.

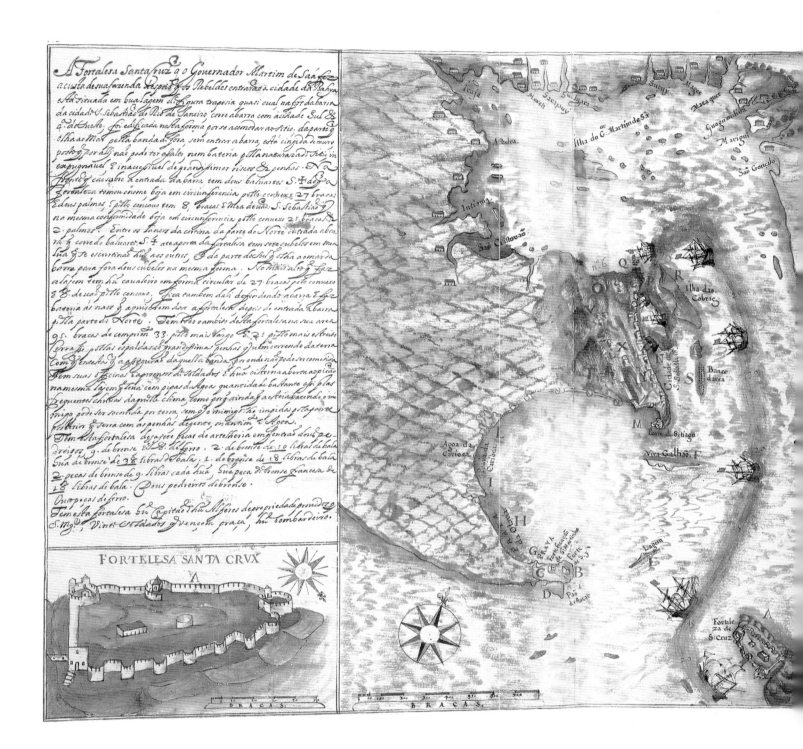

Rio de Janeiro, a manuscript chart by Joao Albernaz, c.1631.
This colourful manuscript chart was drawn in the wake of
several serious Dutch attacks on Portuguese Brazil in the
1620s. Its main purpose is to emphasise the fortifications of
the town of San Sebastian itself, and of Santa Cruz, the fort
which guarded the western entrance to the bay. The
channel into the town harbour is shown, but the overall
geography of the bay is rather inaccurately drawn.

Rio de Janeiro and Brasilia

CAPITANIA DO RIO DE JANEIRO

243

On New Year's Day 1502, a Portuguese squadron under the leadership of the Italian navigator Amerigo Vespucci, which was coasting the shoreline of South America, sailed between two great headlands into a spectacular bay, ribboned by mountains, facing due south, and lying almost exactly on the Tropic of Capricorn. These mariners were in search of a passage around or through the American continent to Cathay and the Indies; they therefore made a rapid survey of the bay, which they took to be the mouth of a river, named the site 'January River', and sailed on. The Portuguese laid claim to this entire territory, but they made no immediate move to settle here, concentrating on northern Brazil, where Salvador (Bahia) was founded and sugar plantations were established. Half a century passed before a second European fleet arrived in the bay, which the Tamoio Indians named Guanabara – 'Arm of the Sea'. This party was French, under Admiral Villegaignon, and they landed on the island which now bears his name. Their purpose was to found a colony to which French Calvinists could escape, and they perversely named the tropical site *La France Antarctique*. By 1560 around 1,500 French settlers had arrived, before the Portuguese awoke to the threat to its overseas empire. A Portuguese fleet set out to claim the bay, but a campaign of several years proved necessary before French colonial ambitions were crushed. It was in 1567 that the Portuguese established Sao Sebastao de Rio de Janeiro, immediately west of where the French island colony had stood. The settlement – the *castello* – was laid out like a medieval citadel, with cannon-guarded walls, and this became the heart of the new city. Sugar cane was planted, and slaves were imported to work the plantations.

For many years Rio was subsidiary to Salvador, but this situation changed at the end of the seventeenth century when rich veins of gold were discovered in Minas Gerais, south-west of Rio. Thousands of prospectors arrived, new roads were built into the hinterland, and the centre of gravity in Brazil shifted south. This was acknowledged in 1763, when Rio was elevated to the status of capital city, in place of Salvador. The real turning point in Rio's history came however in 1808, when the Portuguese monarchy, fleeing Napoleon's invading army, transferred its residence there, and the colonial city became the capital of the Portuguese Empire. The presence of thousands of courtiers, diplomats, soldiers and tradesmen quickly Europeanised Rio and its social life. Schools, shops, offices, theatres, banks, parks and newspapers all sprang up, while international trade raised Rio to the status of a world city. Meanwhile coffee had replaced sugar and gold as the region's source of wealth; planters, exporters and bankers settled in fine houses around the city, and after the return of the king to Portugal in 1821, followed by Brazilian independence in 1822, Rio continued its transformation into a nineteenth-century metropolis, with paved streets, gas lighting, telegraph links and railway lines. The city reached out into the beachside districts to the south and west, Copacabana and Ipanema, accessed through tunnels driven beneath the hills. Streets were widened and old colonial architecture was demolished, as the government aimed at creating a tropical Paris. The first years of the twentieth century saw the birth of the first *favelas*, the unplanned shanty towns climbing the hillsides around Rio, which have in time become cities within the city, as Rio expanded beyond control.

For many years the dominant city of South America, there was yet an awareness that Rio was not Brazil, and that the vast hinterland of the country still remained virtually untouched. This explains an idea that was as old as Brazilian independence itself: the idea of a new capital in the interior, which would shift the nation's centre of gravity westwards, and open up the pampas and the jungle. In the early 1950s a number of sites were surveyed in the highlands, symbolically located among the headwaters of the rivers Parana, Paraguay, Sao Francisco, Araguaia and Tocantins. The building of a new capital was an election pledge of the incoming president Kubitschek in 1955, and his five-year term of office was dominated by the race to complete 'Brasilia' by 1960, while

BRASÍLIA

ESCALA 1:100 000

1 Km 0   1   2   3   4 Km

Rb. Torto

C. Capão Comprido

Rb. Bananal

C. do Régo

Granja
do Torto

C. Urubu

Jardim
Botânico

Bananal

BANANAL

LAGO

C. Poço d'Água

C. do Acampamento

Rb. Bananal

Quartéis

Bosque

Residências
Isoladas Norte

Estação Ferroviária

C. Cana do Reino

Vicente Pires

Residências
Econômicas

Cruzeiro

Setor de Imprensa

Jockey Club

Petrobrás

Iate C

Meteorologia

Setor
Esportivo
Setor Comercial

Estação Rodoviária

Cidade Unive

Esplanada dos Mi

Setor de Indústria

Catedral

Praça do

Estação Abaixadora

Cemitério

Embaixadas

LAGO

Nações

das

GAMA

Jardim
Zoológico

Avenida

C. Guará

Vicente Pires

NOVACAP

DON BOSCO

Vda. da Cruz

Arniqueira ou Roça

NÚCLEO
BANDEIRANTE

Fundo

Aeroporto Comercial

Rb. do Gama

C. Caboclo

C. Mata-Gado

Vda. Grande ou Elias

Riacho

Para
Belo
Horizonte

C. do Cocho

C. Caxiné

Para Anápolis

MANSÕES
SUBURBANAS

C. do Cedro

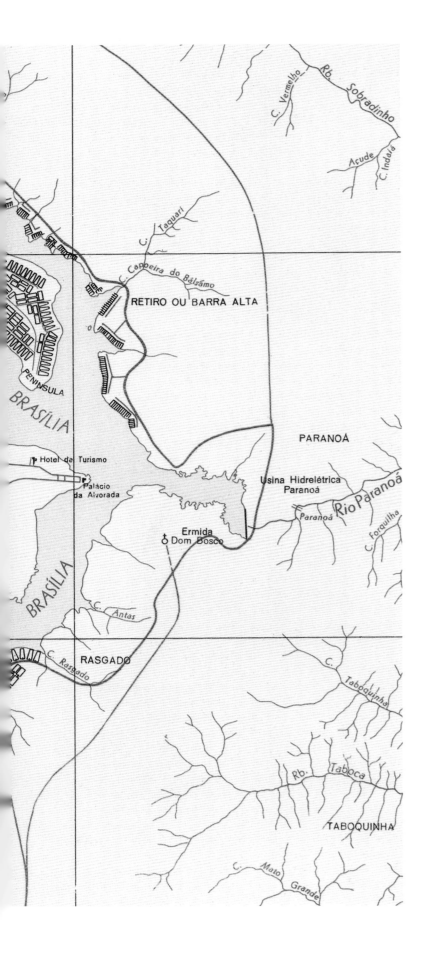

critics poured scorned on the scheme, as a 'dictatorship in the desert'. The overall design of the city was produced and approved in three days, the plan being variously likened to a bird, an aeroplane or a bow and arrow: the body was a monumental axis of government buildings, while the wings were to be the residential quarters. No projections were made of economic activity, population growth, land use, future development or, amazingly, of cost. It was to be a political statement, an architectural symbol of ... what? Of the future presumably, and of the belief that the future can be reshaped in concrete. It was probably the first city in the world built to be accessible only by air; the road from Belo Horizonte, 400 miles (644 km) away, was built later. Some of the individual buildings, designed by Oscar Niemeyer, were admired in their time, especially the part-subterranean cathedral. But within a very few years the regimented concrete blocks came to embody the worst of the dehumanising architecture of the 1950s and 60s.

In fact Brasilia quickly developed a life of its own, but it was in the crevasses of the official structure: a huge, unplanned suburb grew up, centred on the village where the army of construction workers had lived. Taguantina, disliked but tolerated by the authorities, was not a shanty town, but was definitely a concrete slum, home to thousands of small shops and bars, labourers and unskilled service personnel. By 1965, just five years after Brasilia's birth, it was admitted that 100,000 people – one third of the city's population – lived in this zone. The spontaneous growth of Taguantina shows the impossibility of creating a planned city out of nothing, in the face of democratic or market forces. Perhaps only a ruthless, totalitarian regime could achieve it. Brasilia is a place where politicians and administrators have to live, and one from which they regularly have to escape – back to the sprawling informality of Rio.

Brasilia, 1960. One of the many official plans published to celebrate the birth of the new capital. The design – bird, plane or bow? – still looks uniquely imaginative, but the buildings themselves are now sadly dated, and these plans never showed the workers' slums which sprang up immediately on the periphery.
The British Library, Maps 84905 (28)

Rome

The history of Rome is like that of no other city. An Iron Age settlement, no different in 600 BC from scores of others, it became for a thousand years the power-centre of the ancient Mediterranean world. Then, exactly when its secular power was crumbling, it found a new role as the nerve-centre of Christianity, and once again the name of Rome became synonymous with authority, continuity and power. Rome has always been as much an idea as a place, and consequently its name has been revered or hated more than that of any other city. Of the ancient past the visitor now sees comparatively little, for the modern city is largely the creation of the baroque era and after. But the visitor never forgets that he is in Rome, in the city which exercised for so long a unique power over the mind of Europe.

The setting was an unlikely one, for the surrounding land – the *campagna* – was barren marshy heathland, malarial and almost unpopulated for mile upon mile. Here on a group of low hills beside the River Tiber, the early settlements developed into a city-state, ruled briefly by kings, then declared a republic. The inhabitants were Latins, a tribe of unknown origin; they were pastoral people, unphilosophical, inartistic, but practical and warlike, to the extent that their city came in time to dominate the whole of Italy. The city had no special economic base, but flourished by trade and tribute. In the classic process of strengthening its borders and combating its rivals, Rome's territories swelled until they formed an empire reaching from southern Britain to the Caspian Sea. Throughout the classical era, Rome was a city of consumers: soldiers, governors, merchants, craftsmen, servants and slaves, who drew in vast quantities of food and luxuries from all parts of the empire. Tribute revenues were spent on palaces, baths, theatres, circuses and triumphal arches, while aqueducts brought fresh water from the Alban hills, and the Cloaca Maxima carried sewage from the city. By the first century AD the population may have reached one million. From the height of its glory Rome crumbled into decadence from within, and was attacked from without by the peoples of northern Europe. The Empire disintegrated, the centre of

Rome, the ancient city, by Perac, 1574. Visitors and students of Rome have always needed two maps: one of the modern city, in whatever period, and one of the classical city. Tourists and scholars alike have always needed to locate the temples, baths, circuses, mausoleums and monuments among the contemporary streets. Perac's map is a visually impressive reconstruction of ancient Rome, but it embodies a great deal of guesswork: for example, of the Circus Maximus so confidently drawn here just below the centre of the picture, no trace has ever been found.
The British Library, Maps 155 (7)

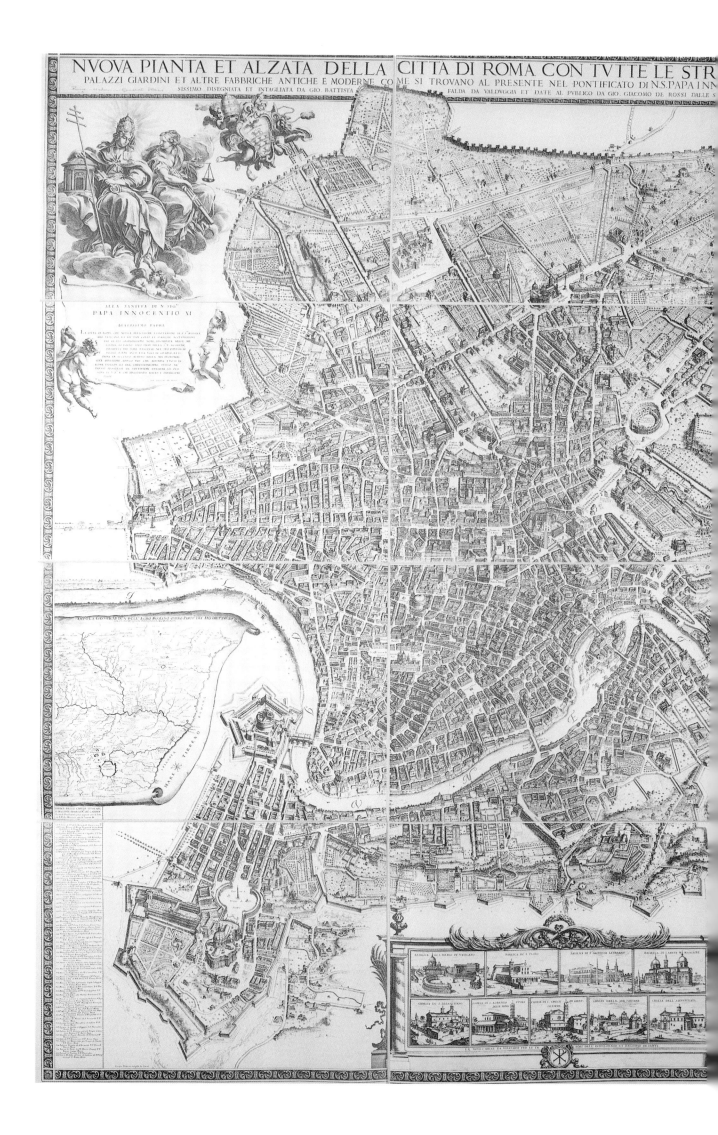

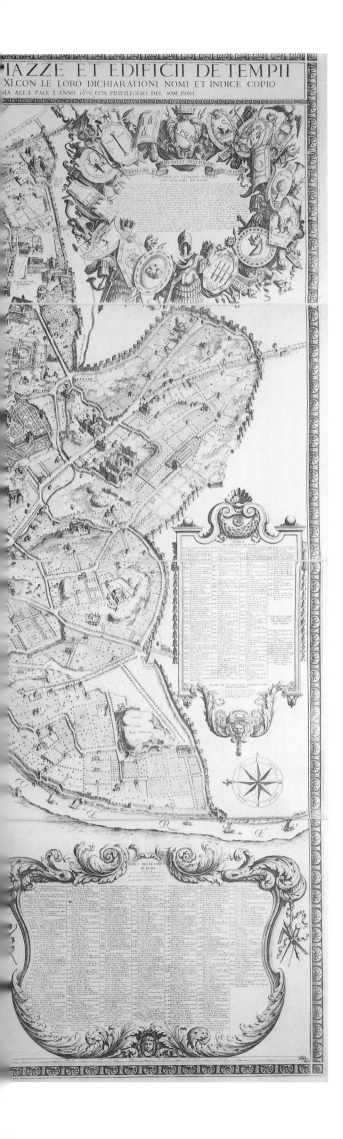

classical civilisation shifted east to Constantinople, and the city became depopulated, an object of contempt or pity.

But a miraculous transformation then took place. By a historical accident Christianity had been displaced from its homeland in Palestine and had taken root in Rome. The bishop of Rome became recognised, in the West at least, as the head of the Christian Church. The power and authority of the city shifted from the temporal to the spiritual realm, and the idea of Rome came to dominate the Christian world as it had once dominated the pagan, classical world. In the process, the pope became the de facto ruler of Rome itself. Relations with Constantinople declined to breaking point, and Rome turned north, forming alliances with the new kingdoms of Europe. When Pope Leo III crowned Charlemagne as Emperor in Rome in the year 800 AD, the idea of Christendom - a Europe united politically and spiritually under a Christian king and pope - was born. This ideal dissolved amid ceaseless conflicts between the pope and the monarchs, and the city itself continued its steady decline. Economically it had nothing to sustain it but the papal court and its many visitors. In 846 invading Muslims plundered the outlying areas of Rome, prompting Pope Leo IV to fortify the basilica and palace - the Vatican, the city within a city. Conflicts within the Church led to the removal of the papacy to Avignon in the fourteenth century. The plague decimated the city, while classical buildings were used as stone quarries. Around 1400 the population was no more than 20,000. Most of the populace became clustered around the basilica of St Peter's, and in the old city the imperial ruins were interspersed with gardens, vineyards or wasteland, where thieves, vermin and wolves prowled by night.

The fifteenth century saw a revival: the papacy returned, and the popes - including two Medicis - reformed the finances of the Church and the government of the city. Roads were paved, squalid houses torn down and new churches built, although in this process the classical remains were plundered still further. Rome took over from Florence the role of leading Renaissance city. Bramanti, Michelangelo, Raphael, Bernini and Borromini fashioned the new city, above all the

Rome, the Falda map, first published in 1676 and reissued with corrections down to 1756. This breathtaking panorama of the baroque city represents the high point of pictorial city mapping, showing all the principal buildings in elevation. In 1748 it was eclipsed for practical purposes by the new Nolli map, in which Rome's streets were shown for the first time in plan and on a consistent scale, without frontal elevations.
The British Library, Maps 3.e.22

new St Peter's on the site of the ancient basilica which had stood for 1,000 years. This process of physical renewal culminated in the reign of Pope Sixtus V, who in the 1580s laid out the street-plan of modern Rome. He commissioned the architect Domenico Fontana to plan new streets and squares and to rebuild many public places. The aqueducts were restored and the fountains remodelled in magnificent sculptural forms. The new Jesuit church of the Gesu inaugurated the baroque style with the soaring depiction of heaven in its dome.

From 1600 to 1800 the Papal State was one of the chief players on the political chessboard of Italy, and in the post-Napoleonic era conflict with emergent Italian nationalism was inevitable, with some of the most dramatic episodes in Rome's long history soon to be played out. Rome itself was excluded from the Kingdom of Italy, founded in 1861, although the Papal States had been annexed. Throughout the 1860s, the Vatican was defended by a French garrison, but when Napoleon III bowed to the inevitable and withdrew these forces, the way was open for Italian troops to enter Rome in September 1870. A plebiscite was held in the following month and Rome was declared the capital of Italy. The arch-conservative Pope Pius IX refused to accept the situation, and declared himself a prisoner in the Vatican. Not until 1929 did the treaty between Mussolini and Pope Pius XI recognise the sovereignty of the pope within the Vatican State. Thus the medieval ideal of a Christian Empire of all Europe had contracted to a few acres within Rome itself. Since 1870 the population has expanded enormously, but despite all the apparatus of government and tourism, Rome stills lacks an industrial base. It is still the idea of Rome which draws people to it, the idea of the city which contributed so much to the identity of Europe, the city governed by emperors and popes, glorified by its artists and hated by its enemies. Rome's history of 2,500 years cannot be said to be unbroken but, as far as any human edifice can be, it is the eternal city.

Rome, St Peter's Square c.1740 by Giovanni Pannini. From the mid seventeenth century onwards, restored, baroque Rome became an irresistible subject for architectural and narrative painters.
Agnew & Sons, London/www.bridgeman.co.uk

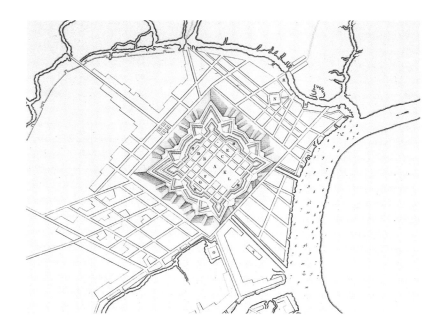

Left: Saigon's fortifications built by the French army in the 1790s while assisting the commander Nguyen Anh, who would eventually unite all Vietnam under his rule. The star-shaped bastions were made on the classic European pattern, and this structure would later form the citadel of the French colonial town.

The British Library, Maps 147.e.119 (14)

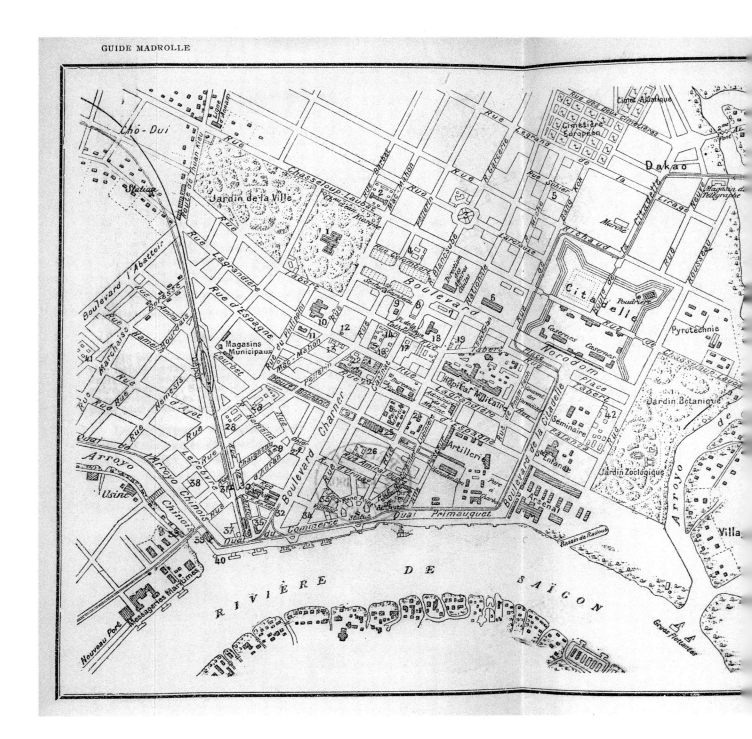

Saigon
(Ho Chi Minh City)

SAIGON

1 Palais du gouverneur général.
2 Collège Chasseloup-Lau-
 bat.
3 Château-a'Eau.
4 Evêché.
5 Poste de police.
6 Hôtel du général.
7 Cercle des officiers.
8 Cathédrale.
9 Presbytère.
10 Palais de Justice.
11 Prison centrale.
12 Commissariat central.
13 Hôtel du Lieut.-Gouver-
 neur.
14 Trésor.
15 Gendarmerie.
16 Enregistrement et Do-
 maine.
17 Recette spéciale.
18 Hôtel des Postes et Télé-
 graphes.
19 Imprimerie coloniale.
20 Hôtel du Secrétaire général.
21 Secrétariat général.
22 Hôtel du Procureur gé-
 néral.
23 Mairie.
24 Manufacture d'opium.
25 Théâtre municipal.
26 Atelier du service local.
27 Justice de paix.
28 Ateliers des Travaux pu-
 blics.
29 Immigration.
30 Poste de Police.
31 Gare de My-tho.
32 Tramways de Cho-lon.
33 Hôtel Ollivier.
34 Poste de Police.
35 Douane.

For ten years from 1965 to 1975 Saigon was transformed from a minor East Asian capital into a city which held the entire world's attention, as it became the focus of the first television war, whose daily horrors were broadcast to millions throughout the world. The fall of Saigon to the communist forces in April 1975 was a televised moment of history that will never be forgotten by those who lived through those years.

The site of Saigon, between the Mekong delta and the South China Sea, was inhabited for almost 1,000 years by the Khmer and Cham people before the Vietnamese occupied the region, although the city itself, then named Prey Nokor, remained in Cham hands until 1690. Inward migration from China and other Asian countries made Prey Nokor a multi-religious city, with Buddhist, Confucian, Hindu and Muslim populations. Christianity arrived in the form of Portuguese missionaries in the 1550s, but later in the seventeenth century it was the French who were particularly drawn to this region, which they called Cochin China. They established trading bases, while French mercenaries fought there in many local wars. It was in 1859-61 that the French, entering into the European race for overseas colonies, took possession of southern Vietnam and of the city which they renamed Saigon. Under French rule, Saigon was developed as an international port exporting rubber, rice and hardwoods. The city was largely rebuilt as a tropical Paris, with wide, tree-lined boulevards and imposing public buildings, including the Cathedral of Notre Dame, the neo-classical governor's residence, the Hotel de Ville, and the Opera House. For almost a century, visitors unanimously described Saigon as having the air of a French city transplanted to Asia and enriched with pagodas and temples.

During World War II the French Vichy regime allowed the Japanese to take control of Indo-China, and Saigon became a Japanese military headquarters. In the 1930s a Vietnamese anti-colonial communist movement had been founded, led by Ho Chi Minh, and with the conclusion of the war, a complex struggle began between nationalists, colonialists and communists. The first phase of this struggle ended in 1954 with French defeat and withdrawal, and with the partition of Vietnam. Saigon became the capital of the southern republic, and awaited the inevitable onslaught from the communist north. When the next phase of the struggle - which we call the Vietnam War - had ended, the destruction was colossal: the dead, injured and dispossessed of the Saigon region alone numbered in their millions. The city was symbolically renamed Ho Chi Minh City, and a period of isolation and rigid communist control followed, during which the city's social and physical fabric was energetically rebuilt. In 1990 the Vietnamese government reopened the country to foreign trade, and, incredible as it would have seemed twenty years ago, the city and the region with its rich Buddhist past is now back on the tourist trail. Today Ho Chi Minh City is a densely populated tropical conurbation, heir to all the problems of both eastern and western cities; one whose history has been violently wrenched out of shape by foreign domination and ideological warfare on a vast and brutal scale.

Left: Saigon, 1902. A deliberately planned European city, set down in equatorial Asia: the orderly grid of streets, the boulevards, the public gardens, the cathedral, the theatre and the palace of justice. The only discordant note appears in the 'opium manufactory', in the very centre of the town.
The British Library, 10055.aa.18

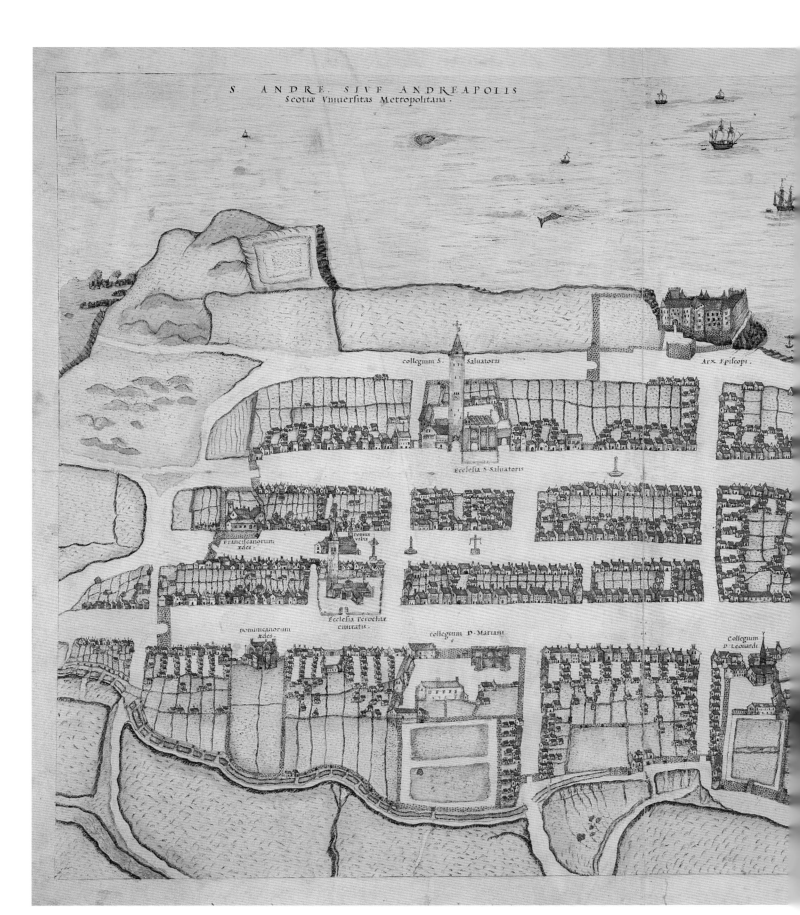

S. ANDRE. SIVE ANDREAPOLIS
Scotiæ Vniuersitas Metropolitana.

collegium S. Saluatoris

Arx Episcopi.

Ecclesia S. Saluatoris

Franciscanorum
ædes.

Domus urbis

Dominicanorum
ædes.

Ecclesia Parochiæ
ciuitatis.

collegium D. Mariani

Collegium
D. Leonardi

St Andrews

'St Andrews by the Northern Sea,
That is a haunted town to me ...'
Andrew Lang

When Andrew Lang arrived in St Andrews as a student in the 1860s, the place was overdue for romanticisation. Built on a ridge of rock whose cliffs fall sheer to the sea, it was a city of ruins, where gaunt towers and arches loomed against the sky, or were dissolved into ghostly indistinctness when the sea mist hung over them. It was the seat of an ancient university, and the gowned scholars who paced its wintry streets could surely have chosen no better place to lose themselves in their studies - a city on the edge of a peninsula, on the way to nowhere, its silence unbroken by trade or traffic or the din of the outside world. It was a place of dreams and books and stones - this was what Lang and many others loved, and embellished in their reminiscences and their poetry. But the silence and the ruins were relatively recent, for St Andrews had once played a major role in Scotland's history, above all during the Reformation, and the ruins of castle and cathedral were the relics of past battles, hatreds and passions. These centuries of fanaticism and violence came finally to an end around the year 1700, and the city entered a period of slumber, undisturbed even by the Jacobite rebellions. The university dwindled to barely two hundred, teachers and students combined; the town reverted to a tiny fishing port, and fragments of the cathedral were carried away by anyone in need of building stone. In 1765 a great storm overwhelmed the tiny fishing fleet, killing most of the crews, and for fifty years even the harbour was deserted. When Dr Johnson visited the town in 1773, he recorded a desolation which 'filled the mind with mournful images'. St Andrews was rescued from this decline in the mid-nineteenth century by the growing vogue for golf and by the arrival of the railway, which transformed the city into a resort town of a rather aristocratic kind.

The map of the town had always been miraculously simple: three long streets radiated westwards from the cathedral, intersected by numerous wynds and alleys at right angles to them. The wynds were narrow and roughly cobbled, as was the central market place, but the two principal streets, North Street and South Street, were wide,

St Andrews, half view, half plan, in a manuscript by John Geddy drawn around 1580, perhaps intended for inclusion in Braun and Hogenberg's atlas of city maps, but not published. The three streets running east-west are still unchanged, while today's 'Scores' was just a clifftop path.

National Library of Scotland, c99/18

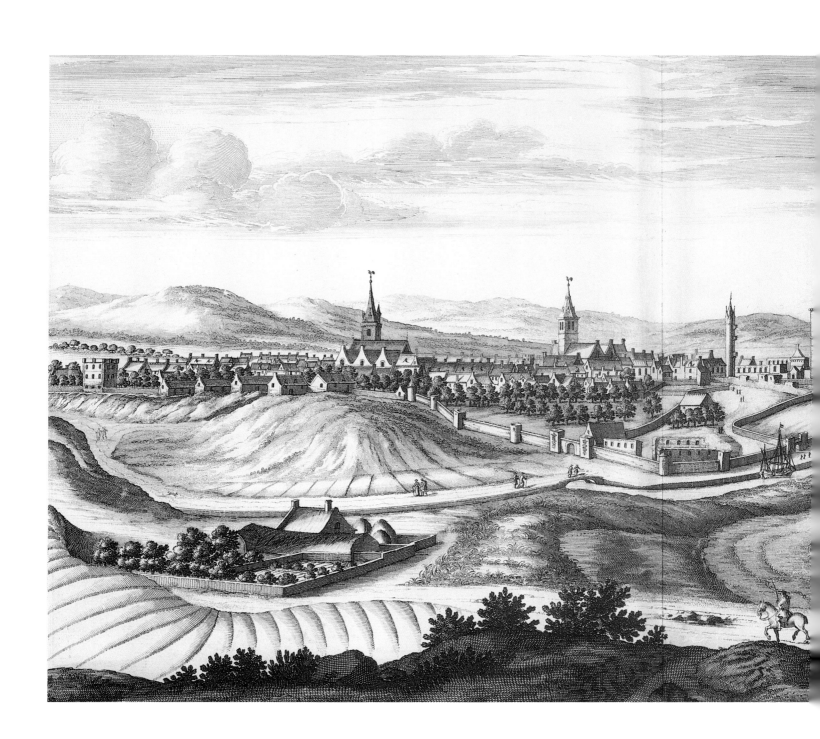

spacious and tree-lined, flanked by the colleges and by fine houses dating from the sixteenth century onwards. The town as a whole was never walled, but the streets were closed with castellated ports, one of which survives on South Street. To the west of the old town, the 1840s saw the construction of several dignified crescents and terraces, in what was almost a miniature copy of Edinburgh's New Town. Because the town has the sea to the north and east, the modern explosion of suburbs and university buildings has all been to the west, leaving the historic eastern core intact. The classic view of St Andrews (since no one now approaches the town from the sea) is still from the hills to the south-east, from the Crail Road, where spires and towers, ruined or living, can be seen rising above the great priory walls, and the rebuilt harbour reaching out into the bay. Lang's romanticism was justified, for this is very much a place haunted by the ghosts of the past, and a place where intellect and passion are concentrated by the physical setting. This feeling is not rare: on any day of the year, one may see a solitary figure wandering the cathedral ruins or the college quadrangles, or gazing into the window of the university bookshop, and sense immediately that this is an old student revisiting the town, revolving many memories, and seeking for something in these streets which he himself could probably not explain.

St Andrews: the classic view of the harbour, the cathedral ruins and the town walls from the hillside to the south-east, from John Slezar's *Theatrum Scotiae* of 1693.
The British Library, 188.f.2

Salt Lake City, 1870, approximately twenty years after its foundation.
The site chosen by Brigham Young extended some three miles by
two on the broad plain below the Wahsatch Mountains. The rigidly
uniform dimensions of the plots and of the streets can be clearly
seen, as can the magnificent temple soaring above them.
Library of Congress

Salt Lake City

Escape was a powerful motive in the founding of the American colonies and cities: escape from corruption or oppressive laws, into an imagined ideal of freedom or righteousness. This motive was present in the founding of many New England colonies in the seventeenth century, when Europe contained the hated regimes. But it was still present in the eighteenth and nineteenth centuries, when the righteous sought to escape from their fellow Americans. The clearest case of all is that of the Mormons, whose distinctive beliefs drove them to found their own community in the deserts of Utah in 1847.

Salt Lake City was in fact the fourth or fifth Mormon city, for, moving from their homeland in upstate New York, they had previously attempted to build their dreams of Zion in Ohio, in Missouri, and in Illinois. Each time they laid out a regular city plan and commenced work on a temple, but each time they were driven away by conflict with their neighbours, in the last of which their prophet-founder, Joseph Smith, was killed in the year 1844. After this last event they determined to trek westward into uninhabited territory where they might find freedom. During this trek, even the settlement which they called 'Winter Quarters' in Nebraska was laid out on a regular pattern. By July of 1847 they had reached the valley of the Salt Lake, far outside any existing jurisdiction, and this was the site selected by the leader, Brigham Young, for the new Zion. Water could be brought by canals from the nearby hills, and the desert would be made to flower. The city was planned on an absolutely rigid squared pattern, with the width of the streets, the building line, and the size of the plots all carefully prescribed. This regularity was considered essential for the setting of a religious community, for the order in the architecture would symbolise law, light and harmony. In this the Mormons were no doubt influenced by the visions of the Holy City in the Old Testament, for example in the Book of Ezekiel.

The first buildings were mostly of adobe, but within three years the population had reached 8,000 or more, and a visitor of 1850 commented: 'The city has been laid out upon a magnificent scale, being nearly three miles in length and two in breadth ... the streets are one hundred and thirty-two feet wide ... the site for the city is most beautiful, for it lies at the western base of the Wahsatch mountains.' Ground was broken for the new temple, and in 1853 the cornerstone was laid, although forty years were to pass before it was completed. The impact of this extraordinary building derives from its proportions, being far higher than its length, so that it appears to soar upwards in an almost impossible fashion. The *New York Times* commented on its completion in 1893: 'Brigham Young conceived the design presumably through the inspiration of some unnatural power, for no such building is to be seen elsewhere in any quarter of the globe.'

The intensely separate identity desired by the Mormons was irreconcilable with the reality of nineteenth-century life and a full half-century of conflict followed the city's founding before Utah compromised sufficiently to be admitted into the Union. The Utopian instinct is a powerful and intriguing one in the history of city building. The temptation to identify the evils of society with the visible evils teeming in the surrounding streets, and to escape both by building a new city, is perennial. It may once have been the stuff of religious dreams, but the modern city planner must still share something of this faith in order to justify his continuing work.

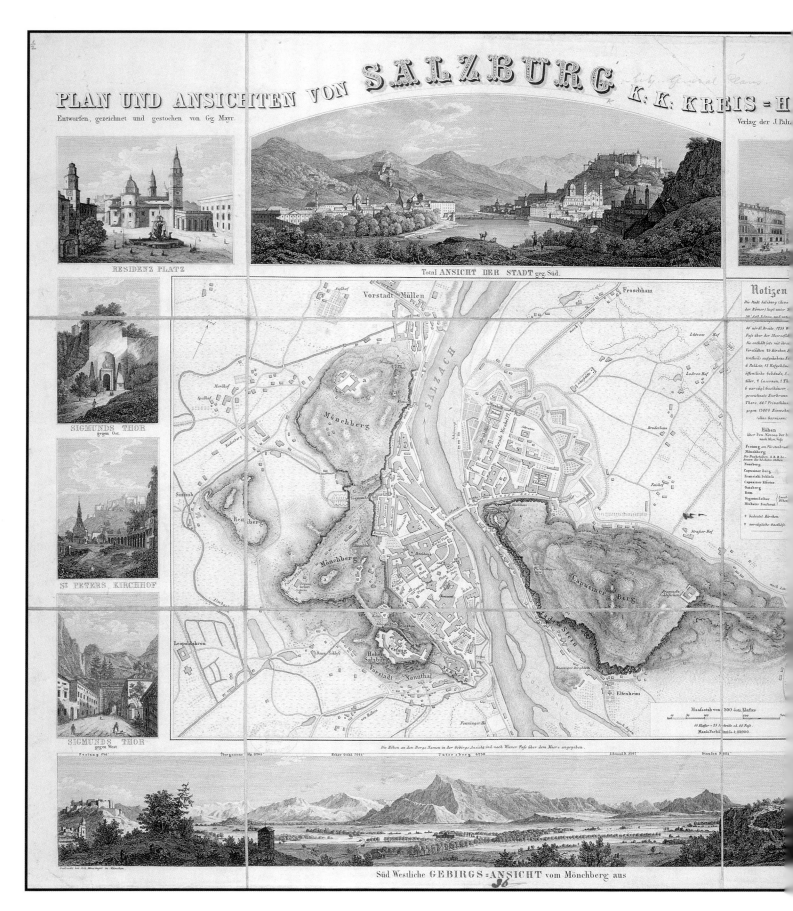

Above: Salzburg in 1841 by Georg Mayr, showing the still compact old city between hill and river. Among the charming vignettes of the city there is no place as yet for the Mozart birthplace, although just one year after this map was published the Mozartplatz would be dedicated, with its statue of the composer.

The British Library, Maps 28327 (11)

Right: Salzburg in 1829, part of a magnificent 360-degree panorama by Sattler and Loos. The hilltop fortress is absent because the view was made from it. The river, the architecture and the hills blend irresistibly together.

Salzburg City Museum

Salzburg

Admired as one of the most beautiful cities in Europe, Salzburg is a fascinating blend of nature and architecture. From a flat plain, steep isolated hills rear themselves above the River Salzach; at a point where the fast-flowing waters run between two such hills, lies the city. Both the hills are crowned with buildings, the southerly one by the massive fortress of Hohensalzburg, while the level ground beneath is crowded with dwellings and churches. Narrow streets open into beautifully proportioned squares, enriched with fountains and statues. The city's natural boundary on the southern side is the ridge-like hill, the Mönchsberg, which offers magnificent views over the roofs, the innumerable church spires and domes, and the fortress itself; to the south the Alps seem but a few miles distant. The whole impression is one of architectural richness and harmony, within a superb natural framework.

Salzburg stands on the northern rim of the mountains, where the trans-Alpine routes emerge into the German plain. From ancient times, the most important commodity traded along these routes was salt, mined extensively from under the nearby hills. A Celtic then a Roman town, Salzburg's modern history began in the eighth century, when it became the seat of an archbishop, who was also a prince of the Holy Roman Empire. Revenues from the salt-mines and dues from mountain traffic made Salzburg the wealthy centre of a self-governing region, ruled until the early nineteenth century by its powerful prince-bishop. Reversing the usual process, the city did not grow around the fortress: rather the fortress was erected in the eleventh century to protect the growing wealth of the community and its ruler. The city was first walled in the thirteenth century,

and the Salzach was bridged to connect the city with the smaller settlement opposite, below the Kapuzinerberg. It was the Prince-Bishop von Raitenau (r.1587–1612), a relative of the Medici family and educated in Rome, who first brought Italianate architecture to Salzburg, creating the first Mirabell Palace. He and his successors ensured that nothing now remains of the medieval city, except perhaps the Steingasse on the north bank. Markus Sittikus (r.1612–19) commissioned the baroque cathedral, modelled closely on St Peter's, and the Palace of Hellbrunn just outside the city. Conquered by Napoleon in 1803, the archbishop's historic powers were stripped away, and city and region were finally integrated within Austria.

It is impossible to think of Salzburg without thinking also of music and of Mozart, the city's most famous son, and the unwitting patron-saint of the modern tourist industry. There is in fact an unmistakable affinity between Mozart's music and the impression produced by this city: the graceful blend of nature and artifice, the alternation of intimacy and grandeur, and the sense of harmony overarching all. Yet, as if in a darker mirror-image of Mozart, the city fathered a second genius who also died in his youth, the poet Georg Trakl, born in the Waagplatz in 1887, and who died in the first weeks of the Great War. Possessed of none of Mozart's felicitous gifts, but a solitary visionary, Trakl's brooding poems seem to have sprung from a different Salzburg, one without people or the heat of summer, a city of cold mists, deserted squares and silvery fountains. Luckily, this other Salzburg is still occasionally to be found on a chance winter's day, while the other Salzburg is a permanent inheritance.

THE CITY OF SAN FRANCISCO.
BIRDS EYE VIEW FROM THE BAY LOOKING SOUTH-WEST.

San Francisco, 1878, by C.R. Parsons. This superb view, from within the bay looking south-west, reveals both the magnificent surroundings of the bay, and the curious way in which the city evolved in three distinct stages by laying out three separate grids of streets, which interlock at unexpected angles.

San Francisco

The headland between the Pacific Ocean and a great inland bay forms the enviable setting for the city of San Francisco. Anywhere in the world such a site would be chosen for settlement, and only historical accident has made San Francisco a young city. Given the superb seaborne entrance to the bay via the Golden Gate, it is perhaps surprising that the site was first seen and explored by land. In November 1769, a party led by Jose Ortega were the first Europeans to look down from the heights of the Santa Cruz Mountains and glimpse the bay. A few years later, in the course of a maritime reconnaissance of the entire coastline, Juan de Ayala first piloted his ship through the Golden Gate in August 1775. A mission and a garrison formed the nucleus of a small settlement, where the Mission and Presidio areas now stand. Development was extremely slow, for the region was isolated both from Spanish Mexico and from the United States. A small harbour village on the north-east side of the headland grew up by the 1830s, named Yerba Buena, where barely two hundred inhabitants dwelled on a neat grid of streets. Whalers and other seafarers recognised the superb potential of the site however, and in 1835 the United States government first tried to buy the bay area from Mexico. One of these early visitors was the writer Richard Dana, who predicted, 'If California ever becomes rich, this bay will be the centre of its prosperity.'

It was in 1848 that the history of San Francisco suddenly accelerated. In that year Mexico ceded California to the United States as part of the settlement of the war begun over Texas, although the town itself had been taken two years earlier and renamed San Francisco. But above all, gold was discovered in the foothills of the Sierra Nevada. Within months, tens of thousands of gold-hungry settlers poured into the town, arriving by sea en route to the hills, or returning to the town for the winter. Tents, plank, mud and adobe houses were mingled together, and goods were piled in the streets for want of buildings to store them. In 1849 one visitor spent a month in the goldfields then returned to San Francisco, and was astonished to find that it seemed to have exploded during his brief absence:

'The town had not only greatly extended its limits, but actually seemed to have doubled its number of dwellings since I left. High up on the hills where I had seen only sand and chaparral, stood clusters of houses; streets which had been merely laid out, were hemmed in with buildings and thronged with people; new warehouses had sprung up on the waterside, and new piers were creeping out towards the shipping. The forest of masts had greatly thickened, and the noise, motion, and bustle of business and labour on all sides were incessant.'

Few of those who came grew rich through gold, but thousands stayed to farm, fish, build ships or trade. The first grid of streets on the north-east of the headland was rapidly extended, and a second grid was set at forty-five degrees to it; these two patterns are clearly seen in plans of the 1860s, and are still traceable today either side of the Market Street boundary line. The bay was rapidly in-filled for new wharfs, and it is said that the bones of many ships abandoned by their gold-rush crews now lie under the streets there. From a figure of 25,000 in 1850 the population approximately doubled every ten years, to reach 200,000 by 1880. The violent growth of the gold-rush years was succeeded by social consolidation: rail-links from the east, churches, schools, a university, theatres, a stock exchange, all supported by wealth from the silver mines of Nevada, from Pacific shipping, from fruit farming and from land speculation, for a golden future seemed assured. The maturing of San Francisco seemed to form a concrete symbol that the United States had fulfilled its destiny of embracing the entire continent of North America, from east to west.

This period was ended by an event still more cataclysmic than the gold rush: the great earthquake of 18 April 1906, a trauma unique to an American city. The subsequent fires raged for four days; four square miles (over ten square km) of the central area were devastated; almost a thousand people died, and tens of thousands of the homeless camped in Golden Gate Park. Rebuilding began at once, and incredibly by 1914, when the city hosted the Panama-Pacific International Exposition marking the opening of the Panama Canal, all traces of the disaster had been wiped out. Years later, geologists discovered that the entire coastline of California lies on the fault line of two great tectonic plates, and inevitably fear of a repeat earthquake will always haunt the city.

San Francisco was long felt to possess a European, and more specifically a Mediterranean, atmosphere, and comparisons with Lisbon were natural. From the late 1960s onwards, the inevitable skyscrapers began to dominate and Manhattanise the city, eroding its uniqueness. Something of the fascination of old San Francisco however was memorably captured in Hitchcock's 1958 film *Vertigo*, where the Spanish past was skilfully counterpointed with the contemporary life of the city.

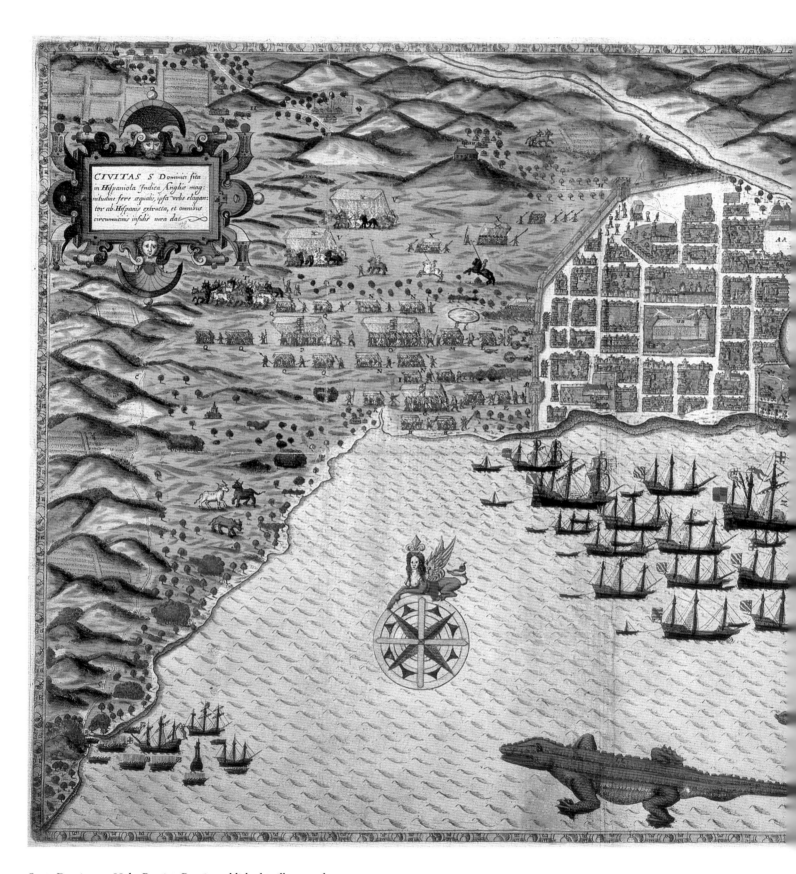

Santo Domingo, 1588, by Baptista Boazio, published to illustrate the account of Sir Francis Drake's West Indian raid of 1585–86, in the course of which Santo Domingo was besieged and captured. Much of the town was looted or burned before the required ransom was forthcoming. The English attacked overland, and their troops can be seen west of the town. The regular grid of streets, typical of Spanish colonial cities, is clearly visible.

New York Public Library, USA/www.bridgeman.co.uk

Santo Domingo

The oldest European city in the New World, Santo Domingo is a direct link with Christopher Columbus, for the city was founded by his brother Bartholomew in 1496, and governed by his son Diego from 1509 to 1516. It was first built on the east bank of the River Ozama, and named Nueva Isabella, in honour of Queen Isabella of Spain, but it was destroyed by a hurricane in 1502, and rebuilt on the river's west bank. Hispaniola was the largest tract of land found by Columbus in the course of his first voyage, and he was impressed by the island's beauty, by the friendliness of the Indians, who called themselves the Taino people, and by the gold and jewels which they wore. The Spanish monarchs awarded Columbus the power to make grants of land on Hispaniola, hence the role of his brother and his son. During these early years, Santo Domingo served as the capital and the starting point for all Spanish enterprises in the New World, and by 1520 the colonists numbered around 3,000. The city church of Santa Maria is the oldest in the western hemisphere, and the university, founded in 1538, is likewise. It was for many years home to Bartolome de las Casas, the great historian of Spanish America and advocate of the rights of the Indians against their ruthless conquerors.

Santo Domingo's subsequent history was largely one of unfulfilled promise. The native Taino people were decimated by diseases brought by the Europeans, and by the end of the sixteenth century were virtually extinct. African slaves were imported to work the land and the mines, with immense and unforeseen consequences for the island's future. The discovery and conquest of the civilisations of Mexico, and later of Peru, shifted the entire centre of gravity of Spanish America westwards, leaving Santo Domingo something of a backwater. It became vulnerable to attack from pirates and from Spain's European rivals, and in 1586 Sir Francis Drake sacked the city, as part of England's unofficial guerrilla war against Spain. A decisive moment in the island's history came in 1697, when Spain ceded the western third of the island to the French who had settled there, thus creating two antagonistic states and a future of inevitable conflict. In 1791 occurred the celebrated slave revolt in French Saint-Domingue (later Haiti) led by Toussaint-Louverture, who then turned his forces against the Spanish colony in the east. In 1801 his soldiers stormed into Santo Domingo and briefly united the island under native rule. A British force helped restore the eastern part of the island to Spain, but in 1821 the colony declared itself independent as the Dominican Republic.

The nineteenth century witnessed a bewildering see-saw of power between the eastern and western states on the island, while Santo Domingo declined into poverty. From 1930 to 1960 the dictator, Rafael Trujillo, turned the nation into his personal estate, and renamed the city Ciudad Trujillo. In the last decade the city has shown signs of recovering from generations of neglect and misrule, with the restoration of some magnificent colonial architecture in the old town. Although Columbus died in Spain, his remains, and those of his son Diego, were taken to Santo Domingo in 1542 and interred in the cathedral there. To celebrate the Columbus quincentenary in 1992, they were moved again to the Columbus lighthouse, but his statue still stands outside the cathedral. The Columbus link has to some extent made Santo Domingo a tourist attraction; but memories of its violent and corrupt past cannot be easily wiped out.

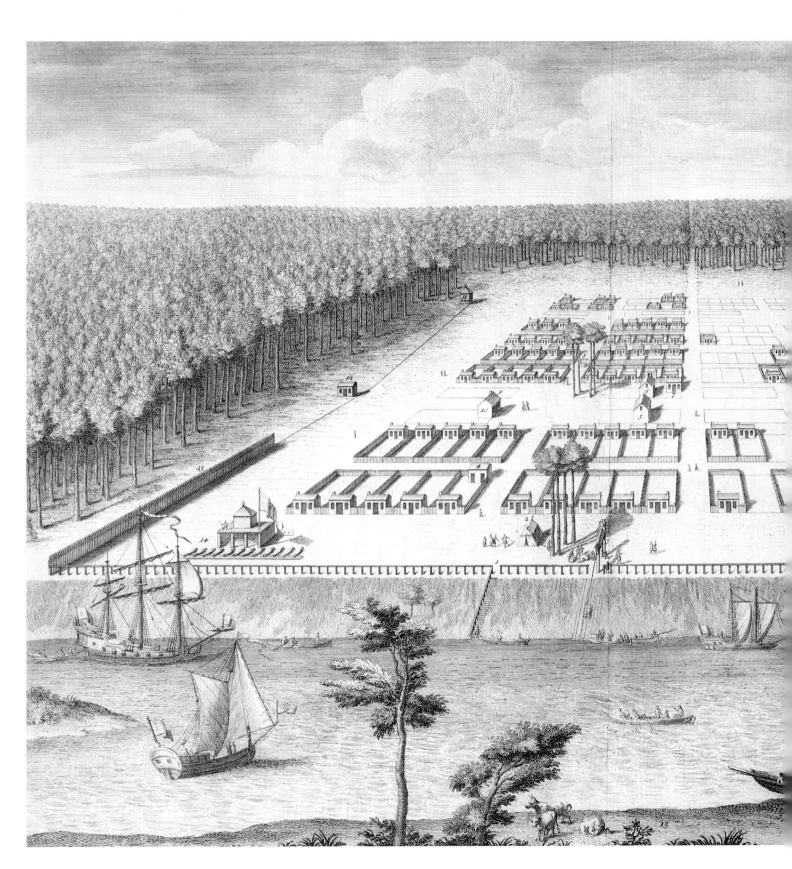

Savannah, by Peter Gordon, 'as it stood on 29 March 1734', that is, a little more than a year after the site had been cleared by Oglethorpe and his English colonists. This is surely the classic image of a planned new town: chequerboard streets and identical house plots emerging starkly from the virgin forest.
The British Library, Maps 1 TAB 44, f.43

Savannah

In the early eighteenth century, the region which later became Georgia was viewed with interest by England as a potential buffer against the Spanish to the south and the French to the east. This presented an opportunity for aspiring colonists, and one appeared in the unusual person of James Oglethorpe (1696-1785), scholar, soldier and philanthropist, who conceived an American colony where the poor, the destitute and the persecuted of England could begin a new life. The colony was chartered by King George II in 1732, and in November of that year 114 suitable people, sponsored by trustees, set sail from London, led by Oglethorpe himself. They arrived at Charleston and very quickly selected a site, for it seems that they had little or no prior knowledge of the area. The chosen site was on a wooded plateau standing embanked 40 feet (12 metres) above the Savannah River. Within two months, trees had been cleared and the first houses built. To prevent land speculation and sub-division of lots, no resale of property was to be permitted, and strict inheritance laws were laid down. Alcohol was forbidden, and an idealistic esprit de corps was encouraged.

The plan for the new city – designed to be around half a mile by a quarter of a mile – was strictly regular and geometric, but it was unusual in that the whole was to be subdivided into six equal neighbourhoods or wards, each with its own large open space, for markets or other public events. The streets were 75 feet (23 metres) wide, running directly south from the riverfront, with others crossing these streets at ninety degrees. The waterfront, called the Bay, quickly became the main place of business, and, much later, of shops and socialising. Savannah largely escaped land speculation, and as it grew the geometric pattern was preserved. From the earliest years, visitors commented on the city as commodious, open, handsome and spacious, a character which it has successfully preserved. The Wesleys came in 1736 to preach to the colonists and the Indians, and Protestant refugees from Europe swelled the population. When war broke out between England and Spain in 1739, Oglethorpe led a vigorous defence of the Savannah region, and the whole town was palisaded. Oglethorpe returned to England in 1743 and pursued a Parliamentary career, never returning to the town which he had so successfully brought to life.

During the struggle for independence, British troops occupied the town, and defended it fiercely in one of the bloodiest battles of the war. After independence, Savannah's importance as a port grew swiftly, and the town became the centre of the world cotton market, supplied by the slave plantations of Georgia. Inevitably, as it was a focus of Confederate forces in the Civil War, it was heavily fortified and became a crucial blockade-running port. Its elimination therefore formed a prime objective to the army of the Union although, having captured it, General Sherman never dreamed of devastating it, but telegraphed to Lincoln on 22 December 1864 that Savannah was his Christmas gift to the President. The perceptive Sherman was indeed charmed with the old city, which struck him as exquisite. By this time its six neighbourhood squares had increased to two dozen, with their shady trees, handsome houses and unmistakable air of elegance. Curiously this urban pattern was little copied in other American towns; perhaps Savannah's rather peripheral status after the Civil War accounts for this. The westward flow of population largely passed the city by, leaving Savannah to cultivate its own special sense of history.

SEVILLE

SEVILLE Ville Archiepiscopale et Capitale du Royaume d'Andalousie
en Espagne; elle étoit nommée ancienement HISPALIS
à Paris chez Daumont

Seville: a view from the south-west across the Guadalquivir.
On the far bank, on the right side from left to right, the vast
cathedral, the Alcazar Palace and the Golden Tower can be
clearly seen, but the network of narrow Moorish streets in
Santa Cruz is invisible here. The ships moored in the river
were the lifeblood of Seville, as the port of entry for all
Spain's trade and treasure from the New World.

The British Library, Maps 18410 (5)

Seville

The capital of Andalusia lies more than 50 miles (80 km) from Spain's Atlantic coast, but the River Guadalquivir, beside which the city is built, has always been easily navigable by smaller vessels. Seville developed into an inland port which was important to the Romans and the Moors, but whose golden age came during the sixteenth and seventeenth centuries, when all the Spanish maritime traffic from the Americas passed through the town. One side of the historic city fronted a loop in the river, while the other side was walled. It was in or near Roman Seville - *Hispalis* - that the Emperors Trajan and Hadrian were both born. In the post-Roman age, Seville was the second city of the Visigothic kingdom of southern Spain, and under its Bishop, St Isidore, it became a notable haven of classical learning. The city was taken by the Moors in 711, who for four centuries made it the leading cultural centre of Muslim Spain, with a population which, if we can believe contemporary reports, reached 400,000 by the twelfth century. The Alcazar, the fortress, is surrounded by the district now known as Santa Cruz, with its characteristic net of narrow Moorish streets and whitewashed houses, where the intense southern sun scarcely penetrates. Indeed from afar Seville resembles an oriental city, with a relatively flat skyline, broken by palm trees and by the *Giralda* - the bell-tower of the cathedral, which was formerly the minaret of the great mosque. The ten-sided tower close by the river - the 'Tower of Gold' - is another great survival from the Moorish period. When the Spanish Christians under King Ferdinand III retook the city in 1248, the cathedral was planned as the largest in medieval Europe, to celebrate the victory over the forces of Islam. The vastness of this cathedral seems to reduce all the surrounding buildings to insignificance. In fact the recapture was a destructive event for the city, for the Moorish and Jewish populations were driven out, the economy collapsed, and a long period of decline followed.

Seville's natural advantage of access to the Atlantic ensured its revival, however. Early in the sixteenth century, it was granted an exclusive royal privilege in trade with New Spain - the Americas. The great *Casa de la Contratacion de la Indias* - the 'House of Commerce with the Indies' - was established here, where the archives of this period still remain, rich with the maps and documents of the Conquistadors. The *Casa* was a hugely important department of state, and its presence raised Seville almost to the level of a royal city. In 1519 Magellan and his crew began their historic first circumnavigation of the globe by sailing down the Guadalquivir to the coast at Sanlucar. Huge wealth in American gold and silver flowed into the city, but the lure of the new world was so great that contemporaries joked that Seville was completely in the hands of women, for all the men had left for America. By the seventeenth century, as ships grew larger, the Atlantic port of Cadiz tended to take more traffic and overshadow Seville, and, as with Lisbon, the local economy was neglected in the rush for overseas wealth. Seville's identity as a city of fashion, theatrical display and relaxed morals provides the setting for two great comedies of manners - *Don Juan* and *Figaro*.

Seville's true architectural richness is the balconies and windows on the small streets and squares, half hidden by flowers throwing out their scents on summer nights, summoning an experience of urban living that is uniquely Andalusian, and scarcely European at all.

23. St Francois.
24. St Isidore.
25. Porte Xerez.
26. Tour d'Or.
27. Tour d'Argent.
28. Porte de Larsenal.
29. le Moulin.
30. Pasluge del Carbon
31. Las Arancanas.
32. Porte du Soleil.
33. Porta de Osario
34. Fauxbourg de Triana.
35. la Riviere de Guadalquivir anciené
36. Porte de Cordoue
37. Porte Triana.
appelée Bœlis

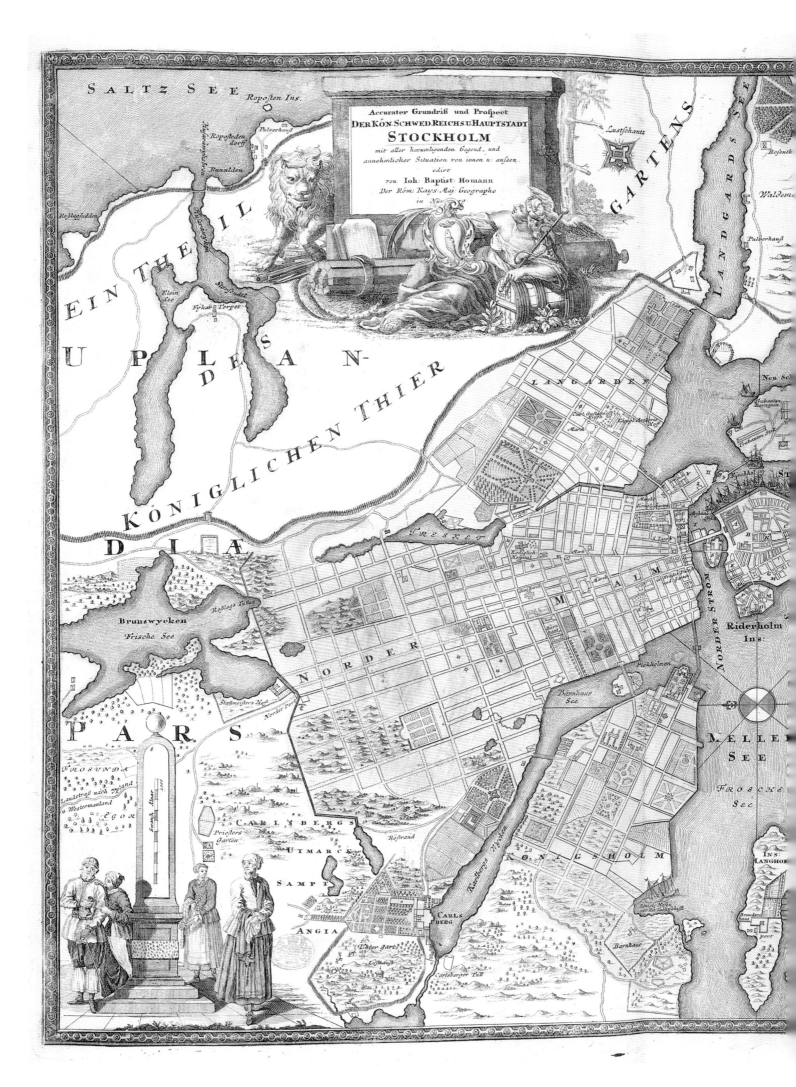

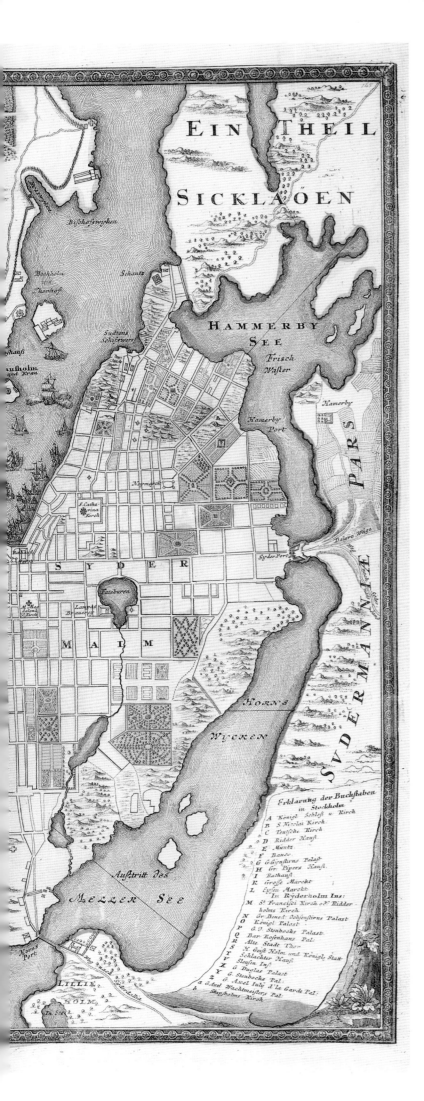

Stockholm

Superbly located on the narrows between a large inland lake and the Baltic, Stockholm is a capital of innumerable islands and bridges, a northern Venice whose waters freeze each winter. Its destiny was for centuries connected with the Baltic, with its trade in fish, timber, metal ores and furs, and the tortuous approach to the city from the sea, through innumerable channels, served as a natural defence. The narrows where the city would be built had been settled since the Iron Age, and the nearby site of Birka was a sizeable Viking town: runic inscriptions of the tenth century tell of expeditions to southern Russia and Constantinople, and westwards to England; one is a memorial to a Viking trader killed in Greece.

Stockholm enters history by name in the thirteenth century, when the central island of Staden was a fortified town of the Hansa League, inhabited by many Germans and trading principally with Lubeck. Throughout the late Middle Ages, Sweden was part of an uneasy union with Denmark and Norway and emerged into full nationhood only in 1523 with the reign of Gustav Vasa, who took Stockholm as his capital. The city at this period was still confined to the island, and was connected to the land north and south by drawbridges. It was fortified by walls and towers, and ringed by a system of quays formed as pilings in the lake. The dominant buildings were the fortress of Tre Kronor, the forerunner of the Royal Palace, and the churches of Storkyrkan and Riddarholm. Gustav Vasa inaugurated the Reformation in Sweden, largely for pragmatic reasons, and in 1527 church lands were confiscated. Building stone from the monastery, the Klara Kloster, went to improve the city's defences, and the seizure of church property immensely strengthened the Vasa dynasty's finances and powers. As settlement spread inevitably to the neighbouring islands, they too were awarded fortifications. For a capital city of some commercial importance, Stockholm remained small: at the end of the sixteenth century, the population was still barely 10,000, but had a sizeable German presence.

It was in the seventeenth century that Sweden became a great power, taking territories in Finland,

Stockholm, c.1720 by J.B. Homann. With the most complex geographical layout of any European capital, Stockholm is spread across a dozen islands, large and small. The strategically placed Staden and Ritterholm formed the historic centre, where the royal fortresses and first churches were built. Only in the eighteenth century, when settlement had spread to Norrmalm and Sodermalm, did the many bridges and waterfronts create the image of the 'Venice of the north'. East is at the top.
The British Library, Maps K. Top. CXI.105

Stockholm, 1768: a beautiful engraved
panorama of the waterfront at the city's heart.
The British Library, Maps K. Top. CXI.107a

Russia, Poland and North Germany, and Stockholm was transformed into the baroque *Residenzstadt* of Gustavus Adolphus and his heirs. Stone houses replaced the older wooden ones, at first in Dutch Renaissance style with ornate gables, then later along neoclassical lines, while a number of churches were rebuilt in the baroque style. The leading architects of this era who set their stamp on Stockholm were the two Tessins, father and son, whose work culminated in the present Royal Palace. There are several Palladian houses to be seen in central Stockholm. This golden period in the city's history ended in the early years of the eighteenth century, with outbreaks of plague which wiped out a third of the population, and a number of disastrous fires, including one in 1697 which destroyed the former Royal Palace. The population had risen in the 1650s to 50,000, but by 1720 it had fallen back to 30,000.

When it came in the nineteenth century, industrialisation made a great impact on Stockholm by breaking down its winter isolation: in the 1820s steamships began plying the Baltic, and in the 1850s a rail line linked the city to Sweden's west coast and to Christiania (Oslo). Early in the twentieth century, ice-breaking ships completed this process by permitting the sea routes to remain open. Modern suburban Stockholm has spread itself irresistibly over an area a hundred times greater than the original area of the island of Staden, but the lakeside heart of the city remains unique in Europe.

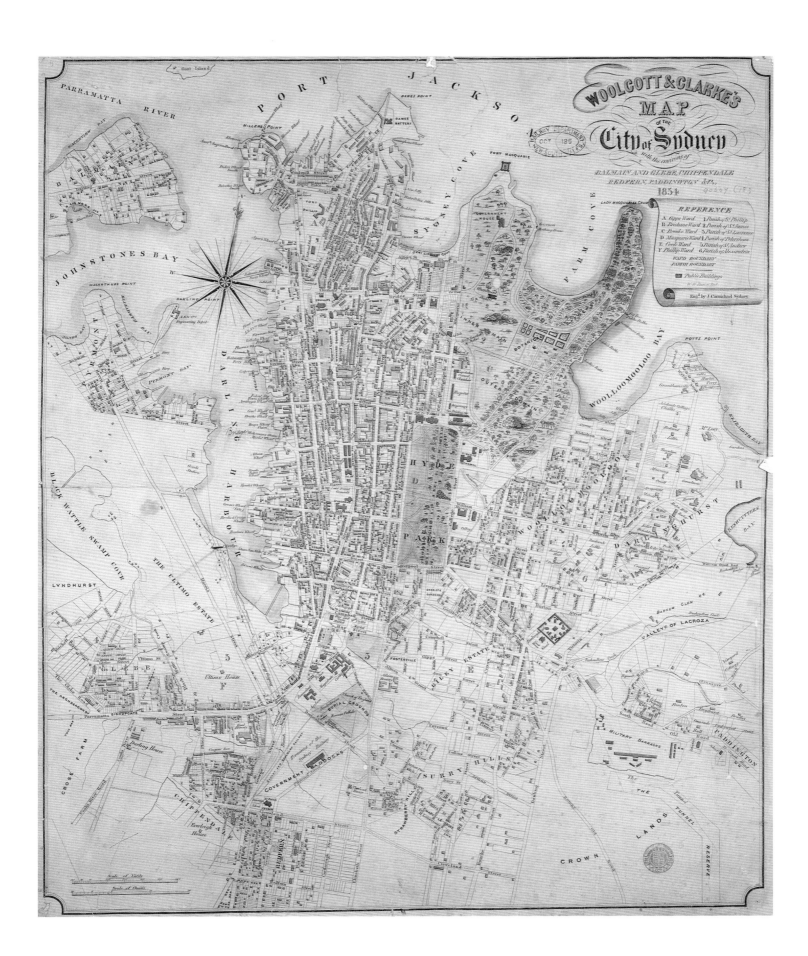

Sydney

Behind Sydney's two hundred years of recorded history lie an astonishing 40,000 years of settlement of the region by aboriginal people. For much of that time the harbour itself did not exist, but was an inland valley which was flooded after the last ice age. The immediate prelude to Sydney's urban history was the arrival on the Australian coast of Captain James Cook in 1770. He entered the wide bay to the south, which he named Botany Bay, for the richness of its flora. Cook and his companions were impressed by the territory, and claimed it for Britain, calling it New South Wales, before sailing away. How the future of Australia might have developed is uncertain, had not the British government then been seeking a site for a new penal colony. Sir Joseph Banks,

Above: Sydney Cove in 1804, drawn by E. Dayes 'from a picture painted at the colony'. Just sixteen years after its foundation the town looks spacious and idyllic in this picture, but the reality at this time was rather different.
The British Library, Maps K. Top. CXVI

Opposite: Sydney in 1854. The British inheritance is apparent in the political and royal street names in the centre: Pitt, Castlereagh, George, Clarence. Darling Harbour penetrated further south than it does today, before it was in-filled; in fact a previous owner of this map has drawn a line for a suggested bridge across its mouth. The newly built railway line appears at the bottom of the map.
The British Library, Maps 90607 (18)

the naturalist who had accompanied Cook, suggested Botany Bay, where, he claimed, convict labour could establish a British foothold in the Pacific region. In January 1788, a fleet of eleven British ships, containing 700 convicts and their guards, anchored in Botany Bay. The leader of the party, Captain Arthur Phillip, quickly realised that the site was less hospitable than Banks had claimed, and he decided to sail on. Just eight miles north he turned into a promising deep-water channel, which opened out into what he described as 'the finest harbour in the world'. At a cove on the southern side, he landed, erected a flagpole and claimed the land once more for Britain, naming the place Sydney, for the Colonial Secretary, Viscount Sydney. Just six days later, two French ships engaged in a reconnaissance of the Pacific under La Perouse also landed at the mouth of the harbour, but left when they saw the British party: how easily Australia might have become a French domain.

The settlement was focused around the creek called Sydney Cove, immediately east of the modern bridge, where international ships now anchor. Conditions for these first settlers were grim: sickness, food shortages, indiscipline, and the hostility of the natives demoralised the leaders and the led. For years, most of the township, with its hovels and its unpaved streets, was an appalling slum, and the only brick buildings of any substance were sited around the governor's house, at the corner of Bridge Street and Phillip Street. The settlement was not laid out in any planned or formal design; rough dwellings and the paths between them appeared anywhere, a legacy still visible in the haphazard run of the streets in central Sydney.

It was the fourth governor, Lachlan Macquarie, who directed Sydney's transition from prison camp to colony. He encouraged the immigration of free settlers, and sponsored the exploration of the hinterland, especially the crossing of the Blue Mountains to the rich pasturage beyond, where sheep farming took root. He employed the gifted architect Francis Greenway, convicted for forgery but freed, who designed Sydney's first public buildings – churches, court house, barracks and banks – as well as private houses. Slum hovels were demolished and streets were widened, and new laws on public hygiene and morals were enforced. When Charles Darwin visited Sydney in 1836 during his *Beagle* voyage, he found it a thoroughly civilised English city, and he concluded that the colony 'had converted vagabonds most useless in one hemisphere, into active citizens in another, thus giving birth to a new and splendid country, a grand centre of civilisation'. Transportation to New South Wales was ended by 1840, and free colonists were being offered assisted passages. The discovery of gold and copper opened a new phase in the history of Sydney (and of Melbourne), swelling the population and stimulating new investment and new building. London was very much the model for nineteenth-century Sydney: the central park was named Hyde Park, the government buildings were modelled on those in Whitehall, the shops imitated the London stores, and the inevitable statue of Queen Victoria was erected in Sydney Cove. The population boomed between 1850 and 1890, rising from 50,000 to 400,000. Few major cities can have had stranger or meaner beginnings than Sydney, but, although never the capital, it has always held its place as Australia's leading city for commerce and culture.

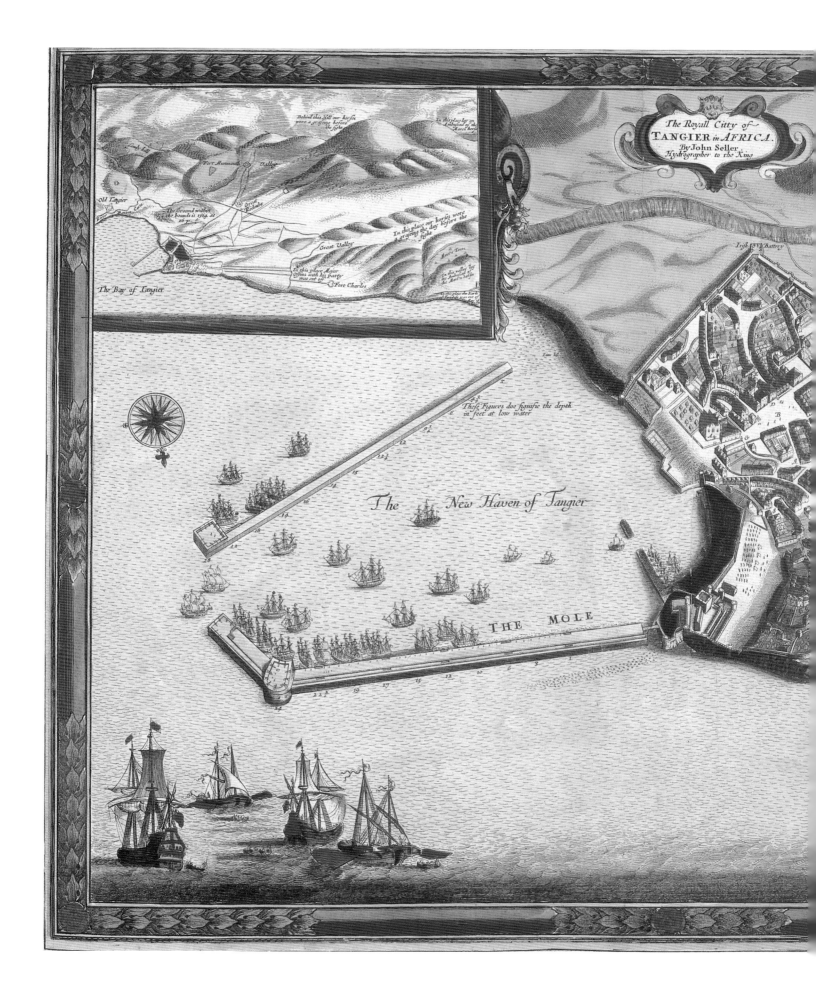

Tangier, *c.*1675, by John Seller. The impressive harbour
had been built by the British, who destroyed it however
when they evacuated the town in 1683. The inset plan
depicts one of the many Moorish assaults on the city,
which eventually made it untenable. The lush green
colouring of this map is quite misleading.
The British Library, Maps 7. TAB. 77 (19)

DIEV ET MON DROIT

A Scale of 100 Yards

Iō · Oliver · Fecit

Governours house

The Upper Castle

Peverbursme Tower

Tangier

For just two decades, from 1662 to 1684, the Moroccan port of Tangier became of intense interest to the British. A Phoenician trading post, then a Roman settlement, Tangier became an Islamic city from the eighth to the fifteenth centuries, until it was occupied by Portugal in the year 1471. In 1662 it formed part of the dowry of Catherine of Braganza, the Portuguese princess who married King Charles II, and it thus became a possession of the British crown; Bombay was part of the same dowry. This acquisition was seen as an opportunity to establish a British naval base in the Mediterranean (Gibraltar did not become British until 1713), and extensive fortifications were built around Tangier, and an impressive harbour was added. However, frequent attacks by Moorish tribesmen constantly undermined the security of the town, and in 1683 the decision was taken to evacuate its British inhabitants and to destroy the new harbour. The diarist Samuel Pepys for many years held the post of Treasurer of Tangier, and he travelled to be present at the evacuation.

This charming view from the north shows the compact ancient port sheltered behind its new fortifications. The colouring is strangely lush, giving the impression that Tangier is surrounded by emerald hills, which is far from the truth. After 1662, the town's streets had been liberally sprinkled with English names – Butcher Row, Salisbury Court, Dean Street and Cannon Street – in a short-lived attempt to impress an English identity upon it. The small inset map gives details of one of the many attacks by the Moors: the country outside the city bears labels such as 'Behind this hill our horse were a-grazing before the fight', and 'In this valley lay the ambuscade of the Moorish horses'. Half a dozen fortified lookout towers were built well outside the town walls, from which signals of any impending attack could be sent. The building of these fortifications and the harbour, and the maintenance of a permanent garrison, proved an unwelcome drain on the royal finances and King Charles was eventually moved to relinquish his strange wedding gift.

John Seller was a leading London chart publisher and hydrographer to the king. He was rather notorious for plagiarising charts, and not updating them, but this town-view was of great topical interest, and it offers an intriguing record of a short-lived British possession on the North African coast. Much later, in the 1920s, Tangier enjoyed a special position as an international city, governed jointly by Britain, France, Portugal, Italy, and, later, the United States. For many years it had a reputation as a raffish and hedonistic resort, home to minor artists, writers, drinkers and worse.

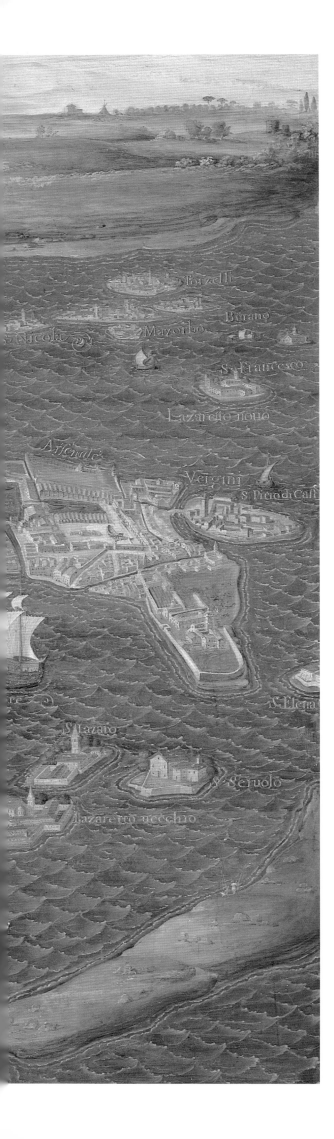

Venice

Venice has long been seen as a work of art rather than as a city. Evoked in prose and poetry, painted and photographed more than any other city in the world, it transcends all its interpreters through its unique blend of stone, water and light. Even in maps and views, its image is like no other, and it has inspired topographical artists to create dramatic views such as this one by Ignazio Danti. Part of Venice's secret is that it has resisted change more than any other great European city: protected by water from the aggressive pressures of population within the city, and from expansion on its edges, its form and character are precisely what they were five centuries ago and more. Unlike the Paris of Renoir or the London of Dickens, the Venice of Canaletto, or even that of Monteverdi, still exists.

Unique among major Italian cities, Venice was not a Roman foundation. The city seems to have originated in the sixth century, with settlements on the sandbanks of the lagoon, perhaps by people seeking to escape the invading Lombards on the mainland. The elected leader of the community - the Doge - enters history by the eighth century, and was soon negotiating political treaties in the name of Venice. The city was situated between the kingdoms of Carolingian Europe and the Byzantine Empire, and this became the source of its trading wealth. By the eleventh century, it had acquired invaluable trading privileges throughout the Byzantine world. Its naval fleet, built in the Arsenal dockyards and maintained at bases in Crete, Rhodes, Cyprus and elsewhere, brought a stream of luxury goods from the east - spices, drugs, silks, gems, gold, furs and artefacts - which were then shipped onwards throughout Europe by land or by sea. On the basis of this wealth, the fabric of the city was enriched with churches, palaces and public buildings. There was another legacy of a grimmer kind however, for Venice was one of the great gateways into Europe for the Black Death, brought unwittingly from Constantinople in 1348.

From the late Middle Ages onwards, the city's governors embarked on the deliberate cultivation of a 'Myth of Venice'. This involved a blend of Christian and pagan beliefs: that the city had been chosen as the resting place of the body of St Mark, but also that it was under the special protection of Neptune. It was ruled by a unique blend of aristocracy and republicanism. Its relationship with the sea was personal and eternal - symbolised in the ceremony in which the Doge 'married' the sea with gold rings each year on Ascension Day. This civic myth was embodied in numerous works of art, in which Neptune or other gods showered the regal figure of Venice with gold.

The rise of Ottoman Turkey in the fifteenth century threatened Venice's position in the Mediterranean, while the discovery of the westward sea routes to Asia after 1500 freed European markets from dependence on Venetian merchants. Economic decline

Venice, by Ignazio Danti, c.1550. The classic view from the south, facing St Mark's and the Doge's Palace. This painting is one of a series of topographical picture maps in the Gallerie delle Carte in the Vatican. It seeks to glorify its subject as in itself a work of art: the sumptuous blue and gold colours reinforce the perception of Venice as a precious, jewel-like city, floating on a rich sea of azure.

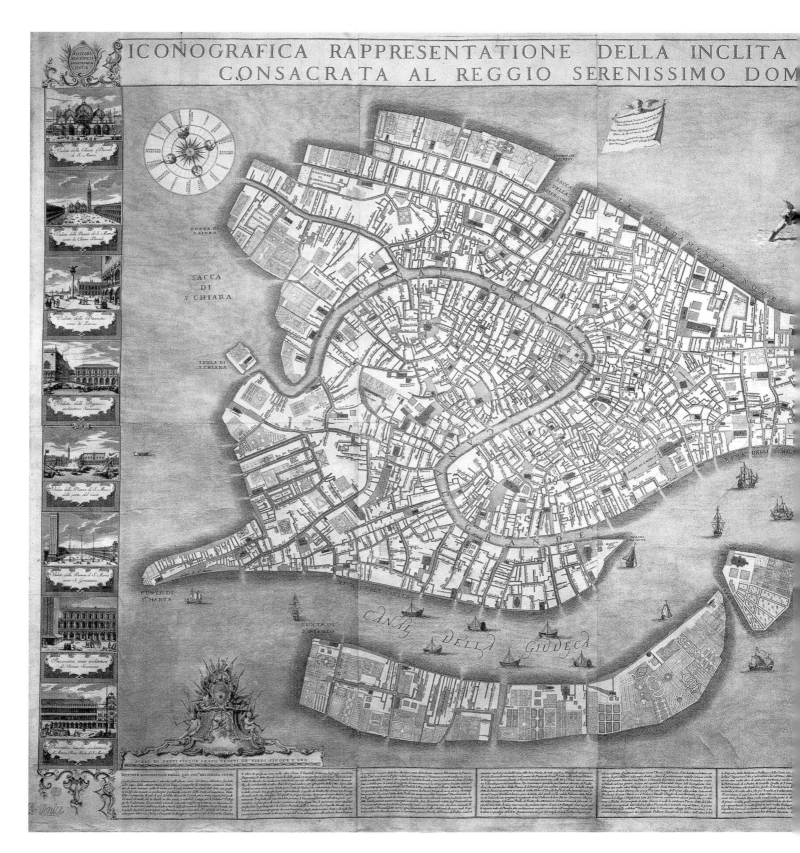

Above: Ughi's map of Venice, 1729, typical of the new generation of eighteenth-century scaled plans which replaced the earlier artistic panoramas. The elevations of the buildings which had formerly adorned the map were transferred to the map's borders.

The British Library, Maps K. Top. 78.63

inevitably followed, and it is easy to find a melancholy note in some of the works of the later Venetian artists, Canaletto and more especially Guardi: the façades of the buildings are neglected and crumbling, the squares are weed-grown, the waters are deeply shadowed, and the human figures insubstantial. In 1797 the Venetian republic was extinguished by Napoleon, and for the next sixty years Venice languished under Austrian rule. In 1866 it was seceded by a weakened Austria to the newly united Italy.

But already, by the mid eighteenth century, Venice had become the first city in Europe to develop a new identity based on tourism. Hotels, cafés, souvenir shops and all the trappings of tourism sprang up, as Venice became an essential destination in any Grand Tour. It was not a city of ideas as Paris was, or of the classical past as Rome was, but it offered an ideal of aesthetic beauty. The architecture, the reflections in the water, the decline from its past greatness, all this offered a sense of a city divorced from time and change, which now existed only to be transformed into art. This sense deepened as the roll-call of those inspired by the city lengthened. After the Italians themselves – Titian and Veronese, Monteverdi and Vivaldi, Canaletto and Tiepolo – there came the outsiders – Turner, Byron, Ruskin, Henry James, Thomas Mann, and many more who came to seek the unique beauty of Venice.

But Venice embodies the classic dilemma of modern tourism: what you come to see is spoiled because every day fifty thousand other people fill the city, seeking the same experience. In 1846 the railway bridge from the mainland was opened, and in 1932 the road arrived. Equally damaging were the corrosive mists from nearby industries in Mestre and Marghera, and the gradual rise of the polluted waters of the lagoon. The floods of November 1966 sounded the alarm, and Venice's physical future is now becoming more secure. But whatever may happen to the stones of Venice, as a place of the imagination its name will always be among the handful of cities in the world which evoke an instant image of beauty, of historic echoes and of an unrepeatable experiment in city building.

Right: Piazza San Marco, a detail from the borders of Ughi's map of Venice.
The British Library, Maps K. Top. 78.63

Vienna: a superb panorama engraved by Nicholas
Visscher, shortly before the Turkish siege of 1683.
We are looking south-west across the Danube. The
overwhelming impression is of the massive fortifications
behind which the population had to shelter. The sense
of living in a fortress at the outpost of Europe, threatened
by alien forces, must have been intense. After the siege,
the spread of the suburbs outside the walls became
irresistible.
The British Library, K90.41.c (1,2, and 3)

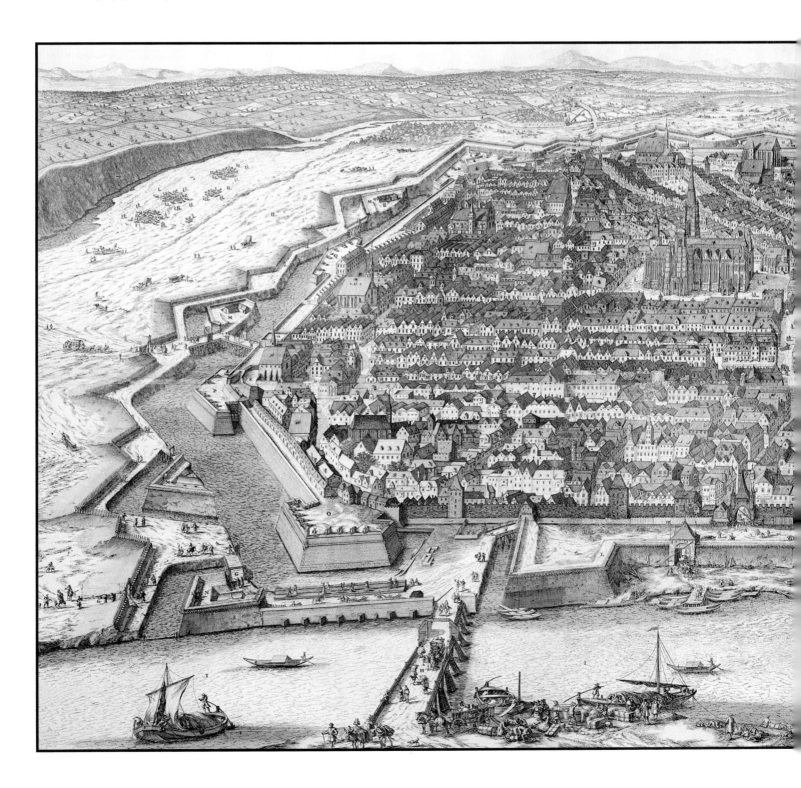

Vienna

For six centuries Vienna was the *Residenzstadt* of the Hapsburgs, and the fortunes of the city were closely tied to that dynasty, as Berlin's were to the Hohenzollerns. Vienna seems always to have been the eastern limit of Carolingian Europe, the 'Eastern realm' or *Oesterreich*. At Vienna, overland trade routes from the Baltic met the River Danube, the great artery of trade with eastern Europe. In the Middle Ages the city was frequently occupied by Hungarians, and on two occasions it was besieged by Turks who were pushing aggressively westwards into Europe. It was after the first siege of 1529 had been beaten off that the characteristic Renaissance fortifications were built to enclose the old city,

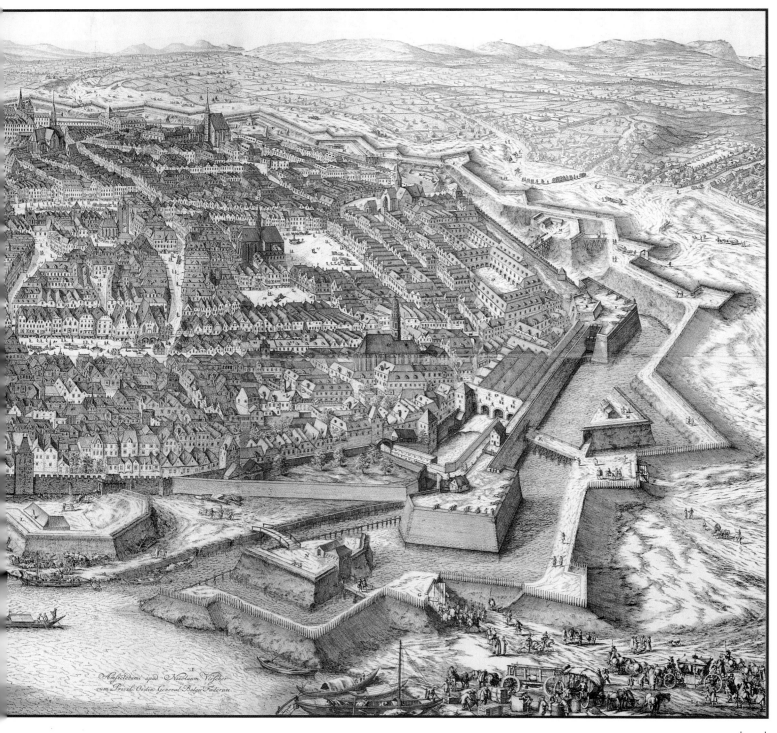

Vienna seen from the gardens of the Belvedere, painting
by Bernardo Bellotto, *c.*1760. The palaces of Belvedere
and Schönbrunn, both well beyond the southern walls
of the old city, represent the post-Turkish siege face of
Vienna, with their rococo display of wealth, elegance
and relaxation.

located where the *Ringstrasse* now stands. These walls were crucial in resisting the second siege of 1683, in which 150,000 Turks were encamped outside the city for almost two months. This conflict was seen by some in Europe as a latter-day crusade, and armies sent from Poland, France, Saxony and Bavaria joined a pitched battle on 12 September which resulted in victory.

The reconstruction after this event was the opportunity to create the baroque Vienna which is so admired today. The Schönbrunn Palace, the Belvedere, the Hofburg and the Karlskirche were the work of Fischer von Erlach, the Wren of Austria, or of his great rival, Lucas von Hildebrandt. With the Turkish threat finally removed, Vienna ceased to be an eastern outpost, and became instead the capital of a large contiguous area including Austria, Bohemia and Hungary. The reigns of Maria Theresa and her son Joseph II (from 1740 to 1790) were enlightened, promoting religious toleration, secularising Vienna's university, encouraging new industry and patronising the arts. Gluck, Haydn and Mozart first made Vienna famous as a city of music, while Beethoven and Schubert would carry this into the new century.

Vienna's centrality made it the natural choice for the post-Napoleonic summit meeting, the Congress of Vienna, in which all the European powers rearranged their borders. This settlement lasted for half a century, until the era of Bismarck. For ten months during the Congress, Vienna was alight with diplomatic energy and social brilliance, orchestrated by the arachnean genius of Metternich. The revolutionary year of 1848 saw street riots in Vienna as in many other European cities, but had no lasting impact; this was the year of the accession of effectively the last Hapsburg Emperor, Franz-Joseph, who was to rule until 1916. In 1857 the great city walls were removed, and in their place boulevards modelled on those of Paris were laid out, lined with public buildings – the opera, the theatre, the museums and the parliament – in a variety of historical styles. Losing a brief war in 1866, Hapsburg Austria finally ceded supremacy in the German-speaking world to Prussia. Yet as in Paris after the catastrophe of 1870–71, there now began a period of outstanding intellectual and artistic brilliance. Music flourished through Strauss and Brahms and then through Bruckner and Mahler; drama through Schnitzler and Hofmannsthal; and painting through Klimt and Schiele; and above all there was Freud. With hindsight, the great question about Freud's theories is whether they truly reveal something permanent in human nature, or simply mirror the abnormal psychology of *fin-de-siècle* Vienna. Rubbing shoulders with Freud in the streets of the city was the young Adolf Hitler (who was influenced by Vienna's anti-Semitic mayor, Karl Lueger), Theodor Herzl, founder of Zionism, and the youthful Wittgenstein.

Hapsburg Vienna collapsed in 1918, when the city was reduced from being the capital of an empire of fifty million subjects to being 'a head without a body', ruling a country of five million, half of them in Vienna itself. There was a certain inevitability about the Anschluss with Germany, and the unanswered question about Vienna, as about Austria as a whole, is whether her role was one of victim of Nazism, a role obviously attractive to Austrians after 1945, or fellow-conspirator? Post-war Vienna is an elegant city which has successfully rebuilt the myth of music and coffee-houses, but the extraordinary personalities and events of the years between 1880 and 1940 suggest a darker current to Viennese history.

Lat. Capitol, 38:55, N.
Long. 0: 0.

GEORGE TOWN

Rock Creek

Presidents House

Capitol

PART OF VIRGINIA WITHIN THE TERRITORY OF COLUMBIA.

P O T O M A K R I V E R.

EASTERN BRANCH

PART OF MARYLAND

OBSERVATIONS
explanatory of the
Plan.

I. THE positions for the different Edifices, and for the
several Squares or Areas of different shapes, as they are laid
down, were first determined on the most advantageous ground,
commanding the most extensive prospects, and the better susceptible
of such improvements, as either use or ornament may hereafter
call for.

II. LINES or Avenues of direct communication have been devised,
to connect the separate and most distant objects with the principal,
and to preserve through the whole a reciprocity of sight at the same time.
Attention has been paid to the passing of those leading Avenues over the
most favorable ground for prospect and convenience.

III. NORTH and South lines intersected by others running due East and
West, make the distribution of the City into Streets, Squares, &c. and those
lines have been so combined as to meet at certain given points with those
divergent Avenues, so as to form on the Spaces "first determined," the different
Squares or Areas.

SCALE OF POLES.

0   100   200   300   400   500   600 Poles.

0   1   2   3   4   5   6 Inches.

Washington

In the doubts and confusions which surrounded its birth, and in the halting fashion in which its final magnificence was realised, Washington seems to symbolise the emergent identity of the new nation in the first century of its life, after 1783. With the War of Independence won, the question of a new capital city naturally arose: given that one was needed, where should it be sited? Sensitivities already existed between northern and southern states which prolonged the argument for a decade, leading to the idea of two capitals, and even to the mocking suggestion of a capital on wheels, which could be pushed around as desired. A site somewhere on the Potomac River was attractive as a compromise mid-point, and in 1790 the newly elected President, George Washington, was delegated to travel the river bank in person and select a site. Of all the places he saw, Washington was most impressed by a wooded section of rising ground just north of the junction of the Potomac with the Anacosta River. It was some three miles from east to west and extended a mile in from the river – spacious enough by eighteenth-century standards for a truly impressive new city. The site was within an area which was designated as the Territory (later the District) of Columbia, a unique arrangement devised because it was felt that no single State should contain the new capital. Equally unusual were the rules that the citizens of Columbia should pay no taxes and have no votes, and be directly governed by Congress itself, rules which were only changed late in the twentieth century.

Washington's personal legacy was not merely the site, but the momentous choice of the planner, a French volunteer officer who had fought in the war, Major Pierre Charles L'Enfant. L'Enfant was the son of a French artist, familiar with the grandiose baroque designs of European courts. He was delighted with the natural rise and fall of the chosen site, and saw a great opportunity to build a city which would become the wonder of the world. The individual plots in the city (whose sale was to finance the project) were designed, as in so many American cities, as a regular gridiron, in this case laid out precisely on a north–south axis. But L'Enfant's stroke of genius was to slash across this gridiron a bold network of diagonal avenues, the three major ones running north-west from the Anacosta River, and these intersected by half a dozen running north-east from the Potomac. There would be vistas everywhere, and from the numerous intersections the avenues would radiate out like stars. The twin focal points would be the presidential palace and the legislative building, the Capitol. It was to be a city of beauty, with gardens, fountains and statues, indeed at the base of the Capitol L'Enfant planned an ambitious man-made waterfall, which however was never built.

Work began in earnest in 1792, but L'Enfant's independent spirit swiftly brought him into conflict with the commissioners who were controlling the project. He was dismissed, and scornfully

Washington, the 1792 plan by Andrew Ellicott, adopted after L'Enfant's stormy dismissal as city planner. More than a dozen principal avenues intersect at unexpected angles, with the aim of creating new vistas at every turn. Perhaps the most grandiose experiment in baroque city planning occurred here in revolutionary America, rather than among Europe's royal capitals.
The British Library, Maps 72310 (1)

Washington in 1892, a century after its foundation: a magnificent
bird's-eye view looking north across the Potomac River, with the
Washington monument in the foreground.
Private collection/www.bridgeman.co.uk

rejected a payment of 500 guineas offered to him. In the end L'Enfant received virtually nothing for his monumental plan, although in 1909 his remains were reburied in the Capitol. His successor was Andrew Ellicott, and although L'Enfant raged that his plan had been 'unmercifully spoiled and altered', in fact it was generally followed quite closely, although building progress was slow. The cornerstone of the Capitol was laid by Washington in September 1793, and work was begun on the White House and on a modest group of buildings to house the administrative offices. In October 1800, government was formally transferred from Philadelphia, and John Adams took up residence in the White House. Even after this, surprised visitors to the capital could still find themselves among woodlands, out of sight of any human habitation. Around 1810 the population still barely 5,000. It was easy to mock the city as a wilderness, and until the arrival of the railways and the telegraph in the 1840s there was a periodic agitation in the press to move the capital to a major city. Washington was temporarily abandoned during the war of 1812–14 with Britain, when British forces burned both the White House and the Capitol.

It was the Civil War between 1861 and 1865 which solidified Washington's identity as the national capital. The government was determined to defend the city at all costs from the nearby Confederate forces. This they achieved, and in the aftermath of the war there began the proliferation of government functions, housed in the neoclassical buildings which now dominate the central area, together with the cultural institutions on the Mall. The suburbs and satellite towns of modern Washington have long outgrown the boundaries of the original District of Columbia, but the heart of the city remains unique in America, a baroque ideal of civic magnificence, a further link between the French enlightenment and American independence, and a series of monuments to the major themes in America's history. Sadly its unique history cannot save Washington from the familiar urban ills of today: crime, pollution or social division, and in these matters, Washington is a city like any other.

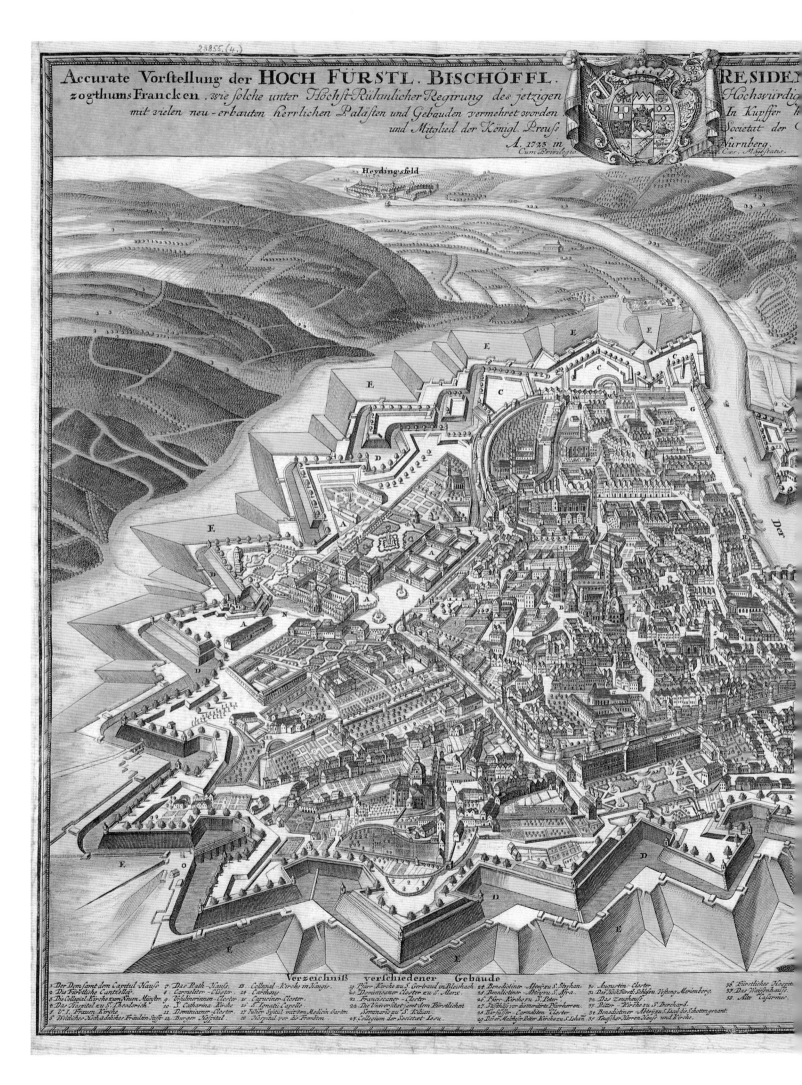

Within the engraving, inscribed text:

d HAUPT~STADT WÜRTZBURG des Her—
sten und Herren Herrn Joh.Phil.Franz Gr. von Schonborn
ben von I.B.Homann. Der Röm. Kais. Maj. Geographo.
affien.

Schloß und Vestung Marienberg

Fluß

hnüß derer neuen Gebaude wie auch derer Thore und vornehmsten Straßen.

Würzburg

Würzburg encapsulates in its surviving buildings much of the history of the central European city: from Dark Age riverside settlement through medieval fortress town, baroque showpiece and war-torn ruin, to urbane modern prosperity. The Altstadt was originally on the east bank of the River Main, where, by 750 AD, there was already a cathedral surrounded by the houses of fishermen and craftsmen and the vintners of what was already a region of noble wine – the Frankenwein. By 1200 the prince-bishops maintained their own army and minted their own coinage, and to mark their temporal power they constructed the fortress Marienberg across the river on a hill which towers 400 feet (122 metres) above the Main, and from here they ruled 'with both crozier and sword'. Besieged during the Peasants' War in 1525, the fortress was enlarged and strengthened throughout the seventeenth century. It was then that the huge bastions with their star-like outworks were built, which still survive, with the hillsides below them clad in vineyards. Similar defences (long since removed) also surrounded the town itself.

But within a few decades of its completion, the fortress had become anachronistic in the new age of elegance, and the prince-bishops commissioned a new palace on the edge of the old town. The Würzburg Residenz, with its frescoes by Tiepolo, is the most magnificent creation of the baroque north of the Alps. Further baroque building flourished in the town. The reign of the prince-bishops was ended in 1802 by Napoleon (who remarked that the Residenz was the prettiest parsonage in Europe), and after the Treaty of Vienna Würzburg was swallowed up in the kingdom of Bavaria.

In a single night in March 1945, British R.A.F. bombers destroyed much of the city. The palace and the fortress were severely damaged, but they have been carefully restored to their former magnificence. The many Renaissance and baroque town houses, especially those fronting the River Main, have gone forever. But, looking westwards from the old Main bridge to the vineyards which lie at the very heart of the town, rising above the slow-moving river to the walls of the old fortress, is still one of the most memorable sights offered by any German city.

Würzburg, 1723, looking east. Würzburg had long been a picturesque city, but the novelty in this engraving was the new palace, the residence of the prince-bishop, visible at the upper left, the most magnificent baroque building north of the Alps. On the southern bank of the River Main, the much older Marienberg fortress tops the steep, vine-clad hill.
The British Library, Maps 28855 (4)

A Plan representing the Form of Setling the Districts, or County Divisions in the Margravate of Azilia.

POTENS ARMIS ATQ: UBERE GLEBA

Azilia and the City of the Sun

unbuilt dream cities

The 'ideal city' has been for centuries a tantalising image that has attracted philosophers, mystics, architects and social dreamers. The ideal has always had two dimensions – that of the physical structure and that of the government – which were, however, always felt to be complementary: order and symmetry in the streets were to inspire social harmony and personal contentment. Beauty and order in the city's physical fabric would foster a collective soul, in which the individual citizen would share. From the Renaissance to the nineteenth century, scores of visionary works were composed in which the shape and the government of the ideal city were set forth. The most influential source of this tradition was Sir Thomas More's

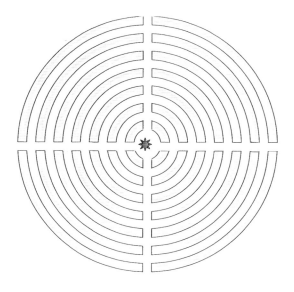

Utopia of 1516, although Utopia was an island and a state rather than a city. More's work was inspired by the age of European discovery, and takes the form of an explorer's account of a civilisation which he has encountered in the New World.

This same device was used in one of the most striking of the post-More Utopias, *The City of the Sun* by Tomasso Campanella, printed in 1626 but composed some twenty years earlier. Campanella's city was said to have been discovered by an explorer in Taprobana, the ancient name for Ceylon. Its physical structure was highly symbolic, for it was nothing less than an image of the Copernican cosmos, consisting of seven great circuits of buildings, representing the seven celestial spheres, at whose centre lay a vast Temple of the Sun. These seven circuits were crossed by four avenues which led from the city's four gates. The Solarians were deists, but not yet Christians, and their religion had a strong dash of mystical science in it, for they believed that the astrological powers of the stars and planets were God's chosen means for directing human destiny. The rulers of the city were adept at drawing down these powers from the heavens, and the conception, birth and upbringing

of all its citizens were carefully calculated with reference to the heavenly bodies. The resulting society was enlightened, egalitarian, and harmonious, while crime, poverty, selfishness and labour were unknown.

The most amazing fact about this fantasy city is its context in Campanella's life: it was composed while he endured long years of imprisonment and brutal torture at the hands of the Inquisition. Campanella was a priest, a mystic and a revolutionary, who attempted to further the cause of freeing southern Italy from Spanish rule. It was in the dungeons of Naples' Castel Nuovo, in the midst of oppression and great personal suffering, that Campanella conceived his city of light as a mirror image of an orderly cosmos, whose harmony he dreamed would one day be reflected in the world of man.

Nearly a century after Campanella, the age of rationalism and colonialism produced a very different and entirely pragmatic plan for a new type of community. This was the 'Margravate of Azilia' conceived in 1717 by the Scottish theorist, Sir Robert Montgomery, and designed to occupy virgin territory between the Savannah and Alatamaha Rivers, where the colony of Georgia, in North America, would later be established. Montgomery's community was to be settled in a precise square of land, twenty miles by twenty, the whole being defended by a strongly fortified perimeter. Since the community was to be wholly self-contained, generous tracts of agricultural land and forest would be included within it. All 116 estates, each exactly one mile square, would be assigned to the colonists and be arranged in an ingenious and entirely symmetrical pattern. At the centre of the little province was to be the capital city, measuring three miles by three, which would be the location of the Margrave's residence, the offices of government and the places of business. In contrast to Campanella's idealistic approach, Montgomery's scheme was purely practical: it makes no direct philosophical point about man's nature, and the orderly layout of Azilia pays homage to a geometric ideal, not a cosmic one. Yet in spite of this, Azilia is still immensely revealing of the Utopian instinct. It suggests the possibility that mankind can wipe clean the social experience of centuries, and begin afresh. Typically of its time, this experiment must be conducted beyond the confines of Europe, with its entrenched political systems, and must be transferred to the virgin forests of the New World. It is significant that Montgomery chose the historical term Margravate: a Margrave was a military governor of a border province, and Azilia was to be a fortress of civilisation in an alien environment. In the event, Montgomery never sold his idea to enough wealthy backers, and Azilia was never built. Montgomery never became America's only Margrave, and so history was deprived of a unique experiment in city planning.

Bibliography

A bibliography covering the individual cities described in this book would be impossible to compile; these are some of the general studies, reference works and picture books which I have found most useful.

Bahn, P.G., *Lost Cities*, Phoenix, London, 1999.

Benevolo, L., *The History of the City*, English Edition, Scolar Press, London, 1980.

Eaton, R., *Ideal Cities*, Thames & Hudson, London, 2002.

Elliott, J., *The City in Maps*, The British Library, London, 1987.

Gutkind, Erwin, *The International History of City Development*, 8 vols, Free Press of Glencoe, New York, 1965-78.

Hall, P., *Cities of Tomorrow*, Blackwell, Oxford, 1988, new edition 2002.

Hyde, R., *Gilded Scenes and Shining Prospects: Panoramic Views of British Towns, 1575-1900*, Yale Centre for British Art, New Haven, 1985.

Mumford, L., *The City in History: its Origins, its Transformations and its Prospects*, 1961, new edition Harcourt, New York, 1989.

Miller, N., *Mapping the City*, Continuum, London, 2003.

Palliser, D.M. et. al., *The Cambridge Urban History of Britain*, 3 vols, Cambridge University Press, Cambridge, 2000.

Reps, J.W., *The Making of Urban America*, Princeton, New York, 1965, new edition 1992.

Ring, T. and Salkin, R., *The International Dictionary of Historic Places*, 5 vols, Fitzroy Dearborn, Chicago, 1995-96).

Toynbee, A., ed. *Cities of Destiny*, Thames & Hudson, London 1967.

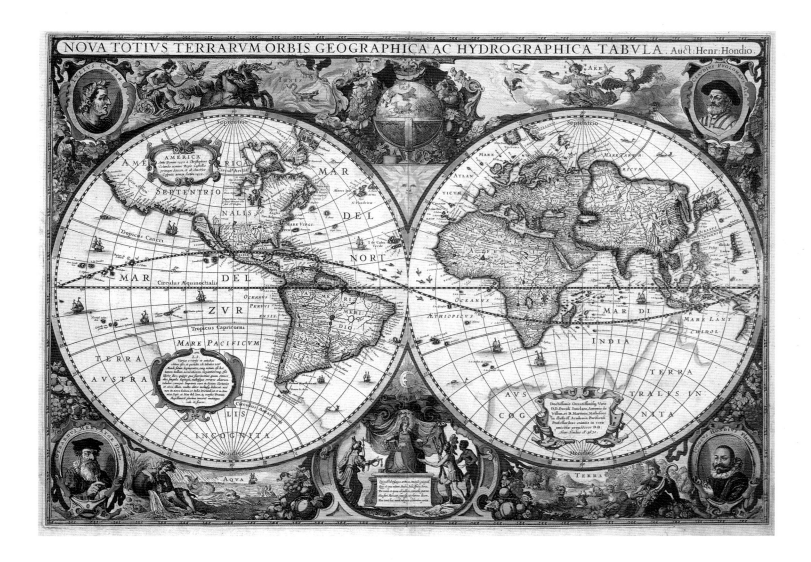

Index